Technology for Humanitarian Action

INTERNATIONAL HUMANITARIAN AFFAIRS SERIES
KEVIN M. CAHILL, M.D., SERIES EDITOR

Technology for Humanitarian Action

Edited by
KEVIN M. CAHILL, M.D.

A Joint Publication of
FORDHAM UNIVERSITY PRESS
and
THE CENTER FOR INTERNATIONAL
HEALTH AND COOPERATION
In Cooperation with
THE STEVENS INSTITUTE
OF TECHNOLOGY
New York • 2005

International Humanitarian Affairs Series, No. 4
ISSN 1541-7409

Library of Congress Cataloging-in-Publication Data

Technology for humanitarian action / edited by Kevin M. Cahill.
       p.   cm.—(International humanitarian affairs ; no. 4)
  Includes bibliographical references and index.
  ISBN 0-8232-2393-0 (hardcover)—ISBN 0-8232-2394-9 (pbk.)
    1. Technology—Social aspects.   2. Disaster relief.
  I. Cahill, Kevin M.   II. Series.

T14.5.T4429   2005
303.48'3—dc22   2004027017

Printed in the United States of America
07   06   05   5   4   3   2   1
First edition

For David and Debbie Owen

Whose creative minds, generous spirits, and transparent integrity are an inspiration for those who must seek new solutions for age-less humanitarian problems

CONTENTS

ACKNOWLEDGMENTS

MANY PEOPLE contributed their valuable time, knowledge, and insight to produce this book. It must be said that the symposium, on which this book is based, would not have been possible without the tireless efforts of President Harold Raveche of the Stevens Institute of Technology and Frank Fernandez, former Director of the Defense Advanced Research Projects Agency (DARPA) and now a Distinguished Professor at the Stevens Institute.

Thanks to Denis Cahill for copyediting and to Chris Comerford for ably assisting him in this undertaking. Fordham University and President Father Joseph McShane generously hosted the symposium, and Robert Oppedesano at Fordham University Press continued his unstinting support for the International Humanitarian Affairs Series of which this book is a part.

A gracious foundation grant to the Institute for International Humanitarian Affairs covered most of the costs of this symposium. The Stevens Institute assisted in the travel expenses.

The authors traveled, some great distances, to speak at the conference and devoted much thought and effort to proposing new solutions for the expanding field of humanitarian assistance.

INTRODUCTION

Kevin M. Cahill, M.D.

HUMANITARIAN ASSISTANCE is a complex undertaking. Even at an individual level the capacity to do more harm than good is ever present. Good intentions are simply not enough, and inappropriate or misplaced aid can further complicate existing tragedies. When whole communities are affected—as happens following floods, famines, earthquakes, and other natural disasters, and, increasingly, during and after armed conflict or as the result of oppression, revolution, political corruption, or incompetence— then often preventable human morbidity and mortality rates rise exponentially.

In previous volumes in this series, colleagues from around the world have considered *Basics of International Humanitarian Missions, Emergency Relief Operations, Traditions, Values, and Humanitarian Assistance* and in a recent volume, *Human Security for All*, have emphasized the inherent dangers and difficult choices of humanitarian work.

Some, if not most, of the solutions to the current limitations for humanitarian workers lie in developing new, appropriate, and cost-effective technology or in discovering how to better utilize existing technology. One of the obvious problems in achieving improved technological application to humanitarian action is in educating the powerful academic, research, and industrial sectors of society that humanitarian assistance is not merely a noble effort to alleviate suffering and provide help to those in need. It is also big business. A recent review article[1] in the prestigious journal *Foreign Affairs* notes that the contribution of private U.S. companies, charitable foundations, religious orders, and individuals is far greater than public government funds. Together, however, they total an impressive $57 billion per year, no mean sum by any standard and surely one that should attract the interest of even those corporations that, understandably, wish to invest in products and programs that can generate profits.

In this book world-renowned scientists consider, many for the first time, how advances in their particular disciplines might benefit humanitarian action. Most of the scientists have worked in defense technology, with close ties to DARPA (Defense Advanced Research Projects Agency). The introductory humanitarian vignettes, written by some of the most experienced workers in the humanitarian field, capture some of the existing problems we face, while the following chapters offer paths to solutions.

Conferences, such as the one that led to this book, expose technology experts to the realities faced on the ground, in the midst of complex humanitarian crises, by workers in constant danger and with inadequate tools to accomplish their tasks. Some current technology is either unknown to (or underappreciated by) humanitarian workers or, even more commonly, has not been adapted for use in refugee camps or conflict zones. Finally, much of existing technology is too expensive, possibly because manufacturers have not understood the scope of the market and that they could rapidly increase production to a level that would significantly reduce the per item cost for new equipment.

Vast sums of money are annually spent to improve military technology, to perfect the weapons, detection, and delivery systems that make modern armies efficient killing machines. What if that technology could be applied to humanitarian efforts? What if the thousands of scientists who helped develop the now defunct Soviet Union's nuclear and other weapons-of-mass-destruction programs could be retrained so that their skills might contribute to helping, rather than annihilating, their fellow man? Could there not be profit in beating our swords into ploughshares?

Technology for
Humanitarian Action

Technology and Humanitarian Actions: A Historical Perspective

Geoffrey W. Clark, Ph.D., Frank L. Fernandez, Ph.D., and Zhen Zhang

WE HAVE CHOSEN at least one humanitarian or human rights effort per decade from the last fifty years to illustrate the relationship between humanitarianism and technology. Our necessarily nondefinitive selection of disasters was based partly on the number of victims or potential victims, partly on the identification of large human needs, and partly on breakthroughs in technologies that have had an impact on humanitarian operations.

We kept in mind the age-old metaphor of the Four Horsemen of the Apocalypse: War, Disease, Famine, and Death. Descriptive historical vignettes start with the specter of nuclear war, nuclear accidents, nuclear terrorism, and the current technological challenge to prepare for radiation cleanup. Three sections of this chapter deal with developments in medicine involving new techniques and their problems, especially given unforeseen threats of pandemics and actual current ones. Several famines are covered, particularly in complex war-torn situations where security for humanitarian or human rights operations was at first lacking and later provided. We also consider the technology of the land mine problem, the inhumane detritus of war and civil war.

The revolution in the news media and satellite communications offers more optimism. Relief organizations have proliferated, and donations have been higher than ever in the last twenty-five years due to new media coverage of disasters. The computer and Internet revolutions have caused encouraging developments, for example in rescue operations and management of humanitarian logistics.

In this chapter we tried to have a healthy skepticism of the relationship between technology and humanitarianism: technologies for war have created the need for counter-technologies for peace, and the threat of terrorist uses of advanced technology has highlighted the need to prepare sophisticated and costly counter-measures.

1. The Specter of Hiroshima and Nagasaki in 1945: Need for international aid in the Chernobyl cleanup, and need for preparation for nuclear threats

Today, in spite of the end of the Cold War, we are haunted by Hiroshima. We still face the threat of nuclear war in several parts of the world, particularly in India and Pakistan over tensions in Kashmir. Fortunately, diplomatic negotiations and the principle of mutually assured destruction successfully prevented nuclear combat during the Cold War and thus assured there was no repetition of the disasters at Hiroshima and Nagasaki.

In the case of Hiroshima, where some 125,000 died immediately and tens of thousands died later of diseases, the U.S. military did not fully realize the effects of radiation on survivors and was unprepared to assist them. There was no attempt at the time to measure the radioactive fallout from the air or determine how to decontaminate affected sites. All the United States did was to give the Red Cross fifteen tons of traditional medical supplies for distribution at emergency hospital units and conduct a joint medical program with the Japanese to track the survivors and study the short- and long-term effects of radiation on the bodies of surviving victims.[1]

This gruesome Japanese study uniquely provides the only human record (as opposed to experiments with animals during nuclear tests) of the physical effects of a nuclear blast should a nuclear war break out—as such it is a humanitarian lesson to mankind and should be dramatized by the media. A nuclear war in central Asia or anywhere else would be an unmitigated disaster for mankind, far greater than anything man-made we have seen before.

Currently, we fear terrorists may attack a nuclear power plant with a plane or some kind of bomb. Even though containment vessels are strong, the result of such an event—or worse, coordinated successful attacks—would call for tracking and predicting the spread of radiation pollution and its cleanup. Although models for the dispersion of radioactive plumes exist, their accuracy depends on environmental variables whose measurement in parts of the world is lacking. Decontamination of radioactive sites is very labor-intensive and cannot, at present, deal with large-scale events.

The experience of the accidental meltdown of Unit 4 at Chernobyl in April 1986 is instructive as a worst-case scenario in such an eventuality. Unlike Hiroshima and Nagasaki, the radiation of some twelve radioactive materials was measured, and their half-lives of up to thirty years were tracked over thousands of kilometers in pollution patterns in the Ukraine and adjacent states. An exclusion zone was established in a thirty-kilometer radius around the unit—the area is littered with radioactive equipment such as trucks and helicopters, and uncharted trenches are full of contaminated soil and other detritus hastily buried after the accident. Western experts have advised that this inner zone will persist as a ghostly wasteland for hundreds of years.

After the downfall of the Soviet Union, whose government had denied the severity of the pollution, the Ukraine established three other zones of heavier to lighter radioactivity and provided compensation to victims in those zones. In the immediate aftermath, 29 people died and 238 suffered from acute radiation syndrome, but hundreds of case-related deaths and disease victims, mostly afflicted with cancers, continued to show up in subsequent years. International standards of permissible exposure were ignored as the Soviets used ad hoc methods to try frantically to stop the meltdown. Subsequently, Ukrainian nuclear management officials tried to institute standards for workers at the site and for permissible levels of radiation in food grown elsewhere in the Ukraine.

The United Nations' International Atomic Energy Agency (IAEA), as well as the industrial world's government agencies and lending institutions, have provided the Ukraine with technical and material assistance to set up new programs. For example, they have assisted the Ukraine in establishing an Information and Emergency Center for nuclear accidents.

"Notifiable" events, classified by international standards, are now reported from Chernobyl and other plants to IAEA as well as to other countries. If the Ukrainian First Deputy Minister decides with expert advice to activate a set emergency plan, selected experts are called in immediately. A data analysis takes place in consultation with IAEA, and an executive group recommends appropriate action. By 1998, on-line radioactivity and meteorological monitoring were established. The plan was to integrate this system with satellite communication and to install real-time, on-line decision-making support from Western Europe.

Unfortunately, the Unit 4 site is still a danger. In 1996 an international group of experts conducted a study of how to make the unit safe. Their recommendations included removal of the fuel, conditioning of the fuel for final disposal, and construction of a confinement building to shore up the disintegrating temporary "shelter" built at the site. The estimated costs are $758 million, and international giving was recommended. Areas of alarm were the corrosion of the lava-like "elephants' feet" made of solid melted sand mixed with fuel in the presence of water: As corrosion proceeds, uranium dioxide is dissolved in water, which may result in the formation of polymeric structures and in a higher concentration of active substances after sedimentation with fragments of the fuel material. This increases the probability of a supercritical mass forming and of a local self supporting chain reaction setting off. An estimated TNT equivalent of such an explosion might be .5 kilograms, enough to release radioactive dust or to destroy the building and containment structures with "more severe consequences."

Negotiations with the Ukraine for aid were, to some extent, held up by the desire of the international agencies to see all the Unit 4-type reactors decommissioned because of design flaws, including the lack of a containment vessel meeting international standards.[2] In April 2002, the UN Undersecretary General for Humanitarian Affairs, Kenzo Oshima, said that Chernobyl was becoming a "forgotten crisis" and presented a ten-year reconstruction and recovery plan that required some $80 million of funding.[3]

There is also a remote possibility of a terrorist group setting off a relatively high-tech nuclear bomb supplied by a so-called rogue state. A more likely possibility is a terrorist group using low-tech, terrorist-made "dirty bombs." Such bombs consist of industrial or medical radioactive materials packed around conventional explosives that, if set off in a city, could contaminate tens of square miles (depending on weather conditions), necessitating extensive evacuation and expensive decontamination—decontamination techniques for types of disasters that we have not experienced before.

It is therefore fitting we begin this chapter with the topic of nuclear war and threats of radiation from accidents and potential terrorism, and to recognize and accept at the outset that developments in technology have rarely been aimed at humanitarian ends. On the contrary, development and proliferation of weapons of war are a major stumbling block to peace and to secure humanitarian operations. Thus, while doing our best through diplomacy and multinational agreements to prevent the worst-case scenarios from taking place, we also have to find new ways to deal with existing weapons, from nuclear ones to land mines, as well as nuclear terrorism and the unintended results of peaceful uses of nuclear power. *Preparation for radiation decontamination is a major area that needs research and development funding.*

2. THE STRUGGLE AGAINST DISEASE: PROSPECT OF REPEATING THE 1957 EFFORT TO MAKE VACCINES DURING A PANDEMIC

Although another influenza outbreak like the pandemic of 1918–19 that killed 20 million worldwide is still possible, we hope it is not as likely due to subsequent advances in medical science and better coordinated world health organizations. It is still possible because the influenza virus, unlike most other viruses, changes its character with each epidemic. However, since the 1918–19 outbreak, pharmaceutical companies have been able to make new vaccines annually to combat current strains of influenza. Such advances, however, do not preclude factory failures that can result, as with the flu vaccine contamination problems in 2004, in national health crises.

Viruses were discovered and grown in the laboratory a few years before the 1918–19 pandemic. Later, with the use of electron microscopes, pharmaceutical companies attempted to manufacture vaccines for specific viral diseases. An early success was a vaccine against yellow fever in the 1930s. The first moderately successful influenza vaccine came on the market in 1945, and others followed to adjust and improve effectiveness.

Then in 1957, an influenza epidemic broke out in Communist

China and was kept secret at first, so there was no immediate reaction. It turned out, however, to be the second deadliest pandemic of the century and, as such, eventually got more attention and organization than even the first pandemic in 1918–19.

In January 1957, the new "Asian flu" spread from China to Japan, Taiwan, Singapore, Malaysia, Indonesia, and India all in one month and caused much alarm. But it wasn't until June that the deaths from the disease were convincing enough to start crash efforts to work on a vaccine. First came efforts to isolate the virus by the U.S. Army Far East Laboratories, and Merck soon developed a vaccine. By July six major U.S. drug companies had developed vaccines and started production. By that time the epidemics had spread throughout southern Asia to Pakistan, to the continent of Australia, and to areas of Europe and the Near East, as well as infecting U.S. military personnel in San Francisco, San Diego, and at air bases in Greenland. The Asian flu had become a pandemic.

As news of the tens of thousands of deaths rose into the hundreds of thousands, panic broke out. It was reported that eucalyptus leaves, thought to be an antidote, were stripped from trees as the disease spread to Colombia on the South American continent. By the fall of 1957, when the pandemic hit the United States, vaccines were still limited and priority lists were established to inoculate crucial personnel, starting with the military and security forces. The U.S. government banned the export of vaccines for sale in other countries. The AFL-CIO called for the inoculation of workers, and the longshoremen's union raised money to try to have its dockworkers inoculated. The World Health Organization (WHO) called for calm, saying that the Asian flu was not as virulent as the 1918–19 strain.

Confirming the WHO's statement, by the time there was enough vaccine in the industrial countries for the mass of the population, the pandemic had run its course with millions becoming sick. Globally, the Asian flu killed 1,250,000 people, the vast majority in poorer undeveloped countries where there were overcrowded conditions in cities and poor nutrition, factors that make influenza deadlier. Only 10,000 people died in the U.S. epidemic.[4]

Clearly, influenza vaccines—unlike those of smallpox, polio, and yellow fever, diseases that have been eradicated or nearly eradicated—are moderately effective in the industrialized world where they are more accessible and more affordable. *One lesson of the 1957 outbreak and experience since then is that the needy populations living in poorer conditions in most developing countries have difficulty obtaining access to vaccines because of high cost.*

The WHO has established the Communicable Disease Surveillance and Response Unit dedicated to the ideal of securing global health by promoting international health regulations—such as more effective government-coordinated quarantine procedures—as well as preparing for epidemics. Besides influenza, this facility also keeps track of new cases of cholera, dysentery, plague, viral hemorrhagic fevers such as Ebola, and many other diseases, including "mad cow" disease, all on databases and accessible via the WHO website.

The WHO and national bodies such as the Centers for Disease Control in the United States have developed early notification and coordinated assistance procedures and continue to play the part of advocates, helping to raise funds for both preparation for medical emergencies and actual outbreaks, especially in poor countries. However, the current estimates for epidemics of influenza are that they will affect 5–15% of the world population yearly with upper respiratory diseases, and typically some 250,000 to a half-million people will die each year, mostly in poor and developing countries.[5] Thus, *without significant increases in economic development and aid from industrial countries, epidemic diseases will continue to bedevil the less fortunate part of the world.*

3. THE STRUGGLE AGAINST DISEASE: WHAT CAN BE DONE IF POLITICAL WILL AND FUNDS ARE APPLIED TO A HUMANITARIAN PROBLEM—THE SUCCESS STORY OF THE ERADICATION OF SMALLPOX, AND CURRENT CONCERNS ABOUT SMALLPOX

Perhaps the most successful example of what can be done if political will and sufficient funds are forthcoming was the campaign of the superpowers and industrial countries to eradicate smallpox from the residual areas in the developing countries where the deadly virus still existed as of the 1960s.

In 1796 Edward Jenner invented a vaccination to inoculate humans against deadly smallpox, an age-old scourge. The disease had periodically ravaged the Eurasian and African continents in epidemics and pandemics, and when Europeans brought small-pox to the Americas during the era of explorations, it rapidly killed an estimated four-fifths of Native American Indians, who had never been exposed to it before. Thus, in 1801 Jenner said of vaccinations, "the annihilation of the Small Pox, the most dread-ful scourge of the human species, must be the final result."[6] The eradication of smallpox was accomplished by 1977. It was the only example of a human disease that has been completely eradicated.

The story of the campaign started when the USSR repeatedly challenged the rest of the world to support eradication of small-pox through the World Health Assembly, the organization that approves the WHO's budgets. By the mid-1960s the Johnson ad-ministration backed the effort, in part to improve relations with the Russians, and the U.S. Public Health Service designated an American doctor, D. A. Henderson, to lead the WHO's Smallpox Eradication Program. The effort with U.S. organization and fi-nancial backing was initiated in 1965 and was successful a dozen years later.

There were skeptics among biologists: for example, René Du-bos, a world-famous microbiologist, expressed grave doubts that smallpox could be completely eradicated from its ecological web. However, microbiologists knew that the variola virus had long ceased to have hosts in nature outside humans, and therein was the hope of eradication. If enough of humanity could be inocu-lated with the vaccinia of cowpox, variola would be driven from its human host. Therefore, Henderson's goal was to inoculate 80% of the most vulnerable of the world's populations, mostly in third world and developing countries, that hitherto could not af-ford the vaccine or the trained personnel to carry out eradication programs.

A breakthrough was made by the pioneering work of another American doctor, W. H. Foege, near the start of the program in 1966. Working in Nigeria to stem an epidemic of smallpox, Foege saw that he was running out of vaccine, and so he tried inoculat-ing people in a ring around the infected population. Eureka! He had discovered the ring vaccination technique of containment. This technique worked so well that it became the key method of "quarantining" the infected with a wall of inoculated people around them. If smallpox broke out of the ring, doctors would frantically seal it up again.

Henderson was credited with being the "Eisenhower" of the war against smallpox, hiring the best people and firing the incompetent, and marshaling the world community—Sweden being a crucial contributor—to put on the payroll some hundreds of thousands of full- and part-time health personnel for the effort. The last battles fought against smallpox, mainly in Bangladesh and Somalia with ring vaccinations, including actions against "containment failures," make dramatic reading. Moreover, even if some of the original motivations had something to do with Cold War politics, *the eradication of smallpox was clearly one of the greatest humanitarian battles and triumphs of all time.*

Even though smallpox has been wiped out from the human body worldwide, the United States has been so fearful of terrorist use of smallpox that officials have started plans for a national program of vaccination. This was partially necessary because both main protagonists in the eradication effort, Russia and the United States, had kept frozen stores of variola virus for several reasons. The benign reason was that reliable experimental stocks were needed at hand just in case smallpox reappeared. The malevolent reason was that variola could be used as a weapon in the Cold War; in fact the Russian military was suspected of manufacturing it as a weapon. After the collapse of the USSR, Western intelligence feared that some of the Russian variola stock might have been sold and might find its way into terrorist hands.

The threat of terrorists became more horrible to contemplate in 2000 when it was learned that two Australian biologists, Ian Jackson and Ronald Ramshaw, had changed the genetic makeup of mouse pox in an effort to sterilize mice and eradicate the pests; shockingly, the injections killed all the pests in the tests. *The new threat is that terrorists could change the genetic makeup of variola or some other communicable disease, a relatively simple thing to do, and have the potential to cause a catastrophic number of deaths.* The old vaccine might not be able to stop such recombinant smallpox. This means that, if such a disaster happened, a vaccine would have to be developed as rapidly as possible as the epidemic (or pandemic) spread, as in the case of the 1957 influenza pandemic, but under much more deadly circumstances.

4. THE STRUGGLE AGAINST DISEASE: NEW EXPENSIVE MEDICAL TECHNOLOGIES AND THE FUNDING NEEDS POSED BY THE HIV/AIDS PANDEMIC

In the past fifteen years more international attention has been given to disasters and humanitarian efforts than ever before through the revolution in computers, the Internet, and wireless telecommunications linked to satellites. Telemedicine was developed and has been used in humanitarian operations in the last two decades:

After the 1985 earthquake in Mexico City, the Advanced Technology-3 communications satellite (ATS-3) provided vocal communication support for the Pan-American Health Organization and the Red Cross. Since all ground-based forms of communication were disrupted, the satellite connection was crucial. ATS-3 gave priority to communications traffic involving disaster assessment and rescue operations.

The U.S./USSR joint Space Bridge Project used telemedicine after the Armenian earthquake of 1988. The joint humanitarian effort provided satellite communications, namely Intelsat and Comsat, to allow linking of four U.S. medical centers with Armenian hospitals for clinical consultation on 209 victims—consultations on cases involving general surgery, neurology, orthopedics, infectious diseases, and psychiatry. The setup included two-way interactive audio and full-motion video.

During the humanitarian intervention in Somalia in 1993, U.S. forces set up the Remote Clinical Communications System using a portable INMARSAT terminal to transmit still, high-resolution, color digital images to the United States for neuroradiological and neurosurgical consultations. In 1994 NASA launched its Advanced Communications Technology Satellite to transmit medical records and live video images quickly and cost-effectively to even remote areas using an ultra-small aperture terminal as a receiver.

Two other methods to lower costs for humanitarian operations are emerging: One way is to make the satellites more powerful in receiving and transmitting so they can pick up and send traffic to small handheld terminals. The other way is to place satellites in a lower earth orbit to communicate with handheld terminals. By 1999, the Global Health Network (GHNet) had been established by the World Bank, WHO, NASA, the U.S. Agency for International Development, and the University of Pittsburgh to link epidemiologists and telecommunications experts. The goals are to foster global health tele-prevention and tele-education.[7]

Such high-tech educational and preventative techniques are needed in the current global HIV/AIDS pandemic. The HIV/AIDS pandemic has the potential to have the same impact on Africa that the Black Death had on Europe in the fourteenth century, when one-third of the population died. AIDS is currently killing millions each year, and since its inception in 1980, 20 million have died. Last year, 5 million people were newly infected. Currently, 40 million people live with HIV, and by 2020 some projections estimate that 500 million will be carriers. Ninety-five percent of those infected now live in poorer developing countries.

HIV is a treatable and eminently preventable disease. The technology is available to stop it, but *the major roadblocks are lack of political will and particularly the lack of aid to the developing countries where the disease is raging.* Another problem seems to be a lack of sufficient media attention to galvanize the industrial populations into providing prevention programs and drugs for treatment, both of which are extremely expensive given the number of people affected.

A ray of hope in the area of treatment has come as humanitarian organizations have pushed governments to take the necessary economic steps to provide the known drugs for HIV *inexpensively* so that the people who need them can obtain them. The patented drugs for antiviral HIV therapy cost up to fifteen thousand dollars per year in the industrial world, thus preventing their widespread use in the developing countries. However, Doctors Without Borders and other humanitarian non-governmental organizations (NGOs) teamed up to put pressure on first-world governments and pharmaceutical companies to amend the World Trade Organization's Trade and Intellectual Property Rights Agreement to make exceptions to meet the needs of poor people who would never be able to pay those kinds of prices. The reinterpretation of the agreement has to be worked out in detail, but *it gives governments in affected countries the right to ensure generic drug manufacturing at lower prices if it is deemed a matter of public health.*

In addition, Doctors Without Borders is currently pressuring first-world governments to allow for free technical transfer of pharmaceutical drug technology to poor countries so that tuberculosis, malaria, and other chronic and epidemic diseases can be treated and contained. In addition, Doctors Without Borders is itself working on new treatments for such widespread chronic diseases as chloroquinine-resistant malaria.[8]

5. THE STRUGGLE AGAINST FAMINE: MEDIA TECHNOLOGY
PROMOTES DONATIONS AND AN INCREASE IN THE NUMBER OF
NGOS DURING ETHIOPIAN FAMINES IN THE 1970S AND 1980S

A chronic threat of famine, or actual famine, lasted for twenty years from the 1960s to the 1980s in drought conditions in western and central Africa, especially in Ethiopia. During the three-year period 1965–68, more than 150,000 Ethiopians died from famine and famine-related diseases such as cholera, severe dysentery, TB, and pneumonia.

There was very little international assistance, and the government of Emperor Haile Selassi did little to generate any aid. In fact, when the drought intensified in 1970–74 and 900,000 starved to death, the government suppressed all news of the seriousness of the new famine conditions and even allowed grain to be exported. In 1973, the government's handling of the famine became a political issue after a student riot in which one student was killed. News of the political opposition and of the famine spread to Europe through diplomatic reports.

Then Jonathan Dimbleboys made a documentary in Ethiopia for BBC television that poignantly depicted the suffering, particularly of a skin-and-bones child opening his mouth in a soundless, starving "silent scream" in his mother's helpless thin arms. This scream was the opening salvo in a media technology offensive against starvation in Ethiopia and, in the long run, an opening to more intense media coverage of humanitarian and human rights causes for the rest of the century. With the galvanizing effect of the film, aid agencies began a partnership with the mass media to raise aid funds. One and a half million dollars was quickly realized so that British charities like Oxfam, Christian Aid, and Save the Children could send food.

Sweden funded a local Ethiopian Nutrition Institute in 1973, and experts in nutrition developed cheap yet healing foods like "faffa" that consisted of 57% grain, 18% soy flour, 10% chickpeas, 8% sugar, 5% skim milk, and 2% vitamins and minerals. The Dunn Nutritional Laboratory in Cambridge, England, developed another successful relief food called "disco." It consisted of vegetable oil, skim milk, and essential vitamins and minerals. Administered to weak, starving children through a tube in several mouthfuls five times a day, the mixture could take a child from near death to walking about in three weeks. Relief camps were set up as shelters and feeding centers, but children continued to die because they were too weak to crawl from hiding places to be fed. Adequate sanitation and potable water were lacking, and children died from diarrhea, pneumonia, bronchitis, typhus, and particularly from epidemic measles.[9]

Meanwhile the Emperor was overthrown in 1974, replaced with the Communist military dictatorship of Mengistu, and a war broke out with Eritrea. The drought continued with greater severity again in 1984 when 300,000 more Ethiopians died and 7 million more were on the edge of starvation.

But in 1984 the media returned with new disseminating tools: for the first time in history, live TV images of starving people in Ethiopia were brought through satellite connections directly into the bedrooms and living rooms of people around the world. The new, more direct media coverage produced "an earthquake in the relief world" and a revolutionary tripling of global aid. In richer countries like the United States, aid donations increased thirtyfold from $11 million in 1983 to $350 million in 1985.[10] *Thus, the 1984 Ethiopian famine was a turning point for the use of mass media to further humanitarian aid—from that point on, television coverage of suffering was considered indispensable for charities.*

In fact, *the mass media and economic boom in the 1980s and 1990s fueled the proliferation and financial growth of humanitarian NGOs.* As a response, NGOs learned to use media coverage to create awareness and raise funds. Such publicity also created pressure on governments to support humanitarian interventions. For example, public pressure resulting from vivid TV coverage of the famine of 1991–92 in Somalia, coupled with a UN plea to assist, led to the decision of the first Bush administration to launch Operation Restore Hope. In addition, haunting TV pictures of women as victims of ethnic cleansing in Bosnia in 1993 spurred the National Organization for Women and other liberal groups to pressure the Clinton administration to finally intervene, something it did not want to do initially. According to an expert on the subject, *"As humanitarian relief workers constantly report, the presence or absence of media attention may mean life or death for affected populations."*[11]

6. The Struggle against War: Humanitarian progress against land mines and the current dilemma

Perhaps no current weapon causes so much suffering as land mines, particularly antipersonnel (APL) mines. APL mines cause an estimated 25,000 deaths globally each year as well as wounding and maiming twice as many. They are relatively ineffective against advancing military units that have the specialized equipment to destroy mines for passage though a minefield, but they are devastating to civilians, mostly women and children.

There are 110 million mines already installed in some sixty-five countries, and 100 million mines are in the arsenals of the world's military powers, insurgent groups, terrorists, and drug-trafficking criminals. In some places in Africa, peasants and farmers cannot till the agriculturally productive areas because of mines and must subsist on outside food aid. For instance, a legacy of the Cold War-inspired civil war in Angola is that the country has 10 to 15 million land mines still in the ground.

Modern plastic APLs are easy and cheap to make, costing as low as three to fifteen dollars. One hundred companies in fifty-five countries make some seven hundred different types of APLs, and unskilled people can easily learn to put them in the ground. However, clearing one APL mine costs an average of three hundred dollars. New plastic mines foil metal detectors, and dogs used to sniff them out sometimes can work effectively for only fifteen minutes. New technologies, including ground-penetrating radar, sound-wave reflections picked up by lasers, robotic clearing devices, and aerial detection of a mine's presence by differences in the color of foliage, have been studied for development, but none has proved as effective yet as the sapper with his handspike—a very slow and dangerous process of clearing mines.[12]

According to UN databases in 1999, the humanitarian challenge to limit APLs was still a priority because 2 to 5 million mines were being laid annually while only 100,000 were being taken out of the ground, and an estimated $33 billion would be needed to clear the already existing minefields.[13]

7. COMPUTER AGE TECHNOLOGY FOR COORDINATION AND MANAGEMENT OF HUMANITARIAN EFFORTS: THE EXAMPLE OF LOGISTICS

A historian of technology, Rudi Volti, once wrote, "It is necessary to add some elements to our definition of technology that go beyond the usual identification of technology with pieces of hardware and ways of manipulating them. The first of these is *organization*. This follows from the fact that the development, production, and employment of particular technologies require a group effort."[14]

Importantly, in the last twenty years the tools of the computer age—satellite and wireless communications, the Internet and websites, management information systems (MIS) and information technology (IT)—have lent themselves to multiple, ongoing attempts at rational prediction of disasters and coordination and management of humanitarian efforts to relieve them.

An early example of UN use of computer databases was in 1992 when the UN created the Humanitarian Early Warning System (HEWS) that was capable of generating data for coordinating humanitarian efforts in more than a hundred countries as well as identifying those with potential crises ahead. Unfortunately, the early system wasn't equipped or connected enough to warn sufficiently about the outbreak of ethnic massacres in Rwanda in 1994, thus pointing out the usual caveat that information systems are only as useful as the data stored in them. Collection of reliable inputs is still problematic, and, in spite of its promise, it would be naïve to think that computer age technology is an automatic panacea.[15]

The struggle to coordinate using new tools may best be illustrated by the example of logistics, so important to humanitarian relief efforts. It also has the virtue of having a limited set of usually numerically important variables as data—something that computers are good at handling. Moreover, modern computerized logistics has been ever more highly developed in the last ten to fifteen years by corporations seeking to increase profitability and competitiveness in the private sector of industrial countries. Logistics has been accepted as part of MIS research and curricula in higher education since it deals with managing and coordinating the flow of information, goods, and finances from suppliers to customers in the most efficient, cost-effective manner.

Humanitarian logistics is unlike commercial logistics, and it is also much more complicated. This might account for why humanitarian organizations have often lagged behind industry in use of up-to-date logistics software and have yet to find a common logistics model. In addition, the range of different aims of NGOs and the range of their activities from goods to services make a common model problematic.

For example, the UN did not create its UN Joint Logistics Center (UNJLC) until 1996 during the eastern Zaire crisis, when there was a need for coordinated pooling of air transportation among UN agencies. Thus, initially the UNJLC had the limited aim of optimizing the use of costly aircraft. The UNJLC was later sent to the Balkans, East Timor, Mozambique, Angola, Afghanistan, and, in 2003, to Iraq. In these places, the organization became involved in representing the community of humanitarian organizations in negotiations with governments for access to facilities and with suppliers for joint commodity prices.

Examples of the logistics coordination work of the UNJLC are found during the war in Afghanistan in 2002. The UNJLC set up a website for the crisis that contained information for logistics personnel of government agencies and NGOs. With feedback, the website was updated and made more comprehensive to reflect planned regional and strategic airlifts, status of transportation corridors, availability of warehousing, rates for commodities and transport, and details like the status of border crossings. The UNJLC also addressed logistics bottlenecks as a result of the actions of neighboring states. For example, the main land route for humanitarian aid in food, medical, and sanitation supplies for displaced refugees was through the Uzbek corridor in the north, where there were port facilities, rail lines, and asphalted roads on each side of the Amu Darya river bridge. However, the local Uzbek authorities had closed the bridge for security reasons several years before, and, in the chaotic aftermath of the war against the Taliban, they resisted requests to open it to humanitarian traffic. A UNJLC team met personally with the local authorities and negotiated an opening of the bridge so that surface transport resumed between the two countries for the first time since 1998.

Another problem in Afghanistan was the tendency of humanitarian organizations to compete for the same facilities and drive up prices. For example, in Herat NGOs bid up the price of trucks 300% in a six-month period. Millions of dollars were saved when the UNJLC negotiated with the Herat trucker cartel for a set reasonable transport price for all organizations. They won the concession by threatening to import a UN fleet of trucks if the truckers refused to agree, and UNJLC placed the agreed-upon price on its website and thus stopped the gouging of humanitarian agencies.

In addition, in Afghanistan, the UNJLC produced assessments of roads for proposals to donors for reconstruction of the road network and, in both Afghanistan and Iraq, negotiated with Coalition forces for the speedy opening of airfields for humanitarian flights. During these efforts, the UNJLC worked on the principle that each humanitarian organization's logistics had to be respected, and therefore the UN did not attempt to standardize logistics for all humanitarian organizations.[16]

In fact, standardization may not be desirable in all cases. For instance, some logistics operations in complex emergencies depend more on intangibles for their success than on efficient software and expert managers trained to use it correctly. It often seems that the dedication of the NGOs personnel, their knowledge of the people they deal with at the local level, and their bravery in the hostile environment are more important than having the latest software. In fact, they often use a simple stock card system that can be kept in the pocket, only transferring it later to a computer for tracking and record-keeping purposes.[17]

However, logistics experts say many NGOs need to catch up on the latest logistics technologies as they are using techniques that the corporate sector was using ten to fifteen years ago. NGOs have spent little capital on implementing up-to-date management information systems, information technology, or logistics software systems.[18] Moreover, information technology is used in widely different ways among NGOs, and real-time logistics information is not readily available to them.

Experts from the Fritz Institute, a philanthropic organization dedicated to assisting humanitarian organizations to improve their logistics capabilities, point out the following deficiencies in relief organizations' logistics systems: multiple spreadsheets, inadequate budget control, untracked usage of funds, procurement procedures that are difficult to enforce, manual tracing of shipments, no central database for a history of prices paid or transit times, and reports that are done manually. They call for use of "commercial best practices . . . adapted to humanitarian requirements through extensive research with many leading relief organizations."

The Fritz Institute donated more than a million dollars and three thousand man-hours in a multiyear effort to assess the International Federation of Red Cross and Red Crescent's (IFRC) logistics technology requirements. The result was a software package, the Humanitarian Logistics Software, developed by the Fritz Institute for the IFRC and its national members' network. This software was put into operation by the Red Cross system in September 2003.

The software consists of a web-based package that mobilizes people, resources, skills, and knowledge to address disasters effectively. Its four modules are mobilization, procurement, transportation, and tracking: "The mobilization module simultaneously tracks needs of the beneficiaries and agency funding appeals, reconciling them with donations. The procurement module controls purchase orders, performs competitive bid analysis and reconciles received goods against invoices awaiting payment. The transportation and tracking module allows consolidation of supplies for transportation and allows the automatic tracking of major milestones in this process."[19] As data is built up over time, some of the modules produce information on the performance of suppliers and transportation vendors, and a reporting module provides detailed reports for donors and decision makers. In particular, in the case of food, medicine, and shelter, the program keeps an up-to-date accounting of appeals for donations, donation figures, procurement, transport, tracking, customs clearance, local transport, warehousing, and the last mile of delivery to disaster victims. It promises to automate and standardize the relief mobilization process and provide more control and efficiency.[20]

CONCLUSION

For the future, there is a lot of potential for adapting and creating technologies for humanitarian ends, but new technologies will not automatically be put to humane uses without the political will and the economic means to do so. This necessitates building upon and furthering the recent trend of enlargement of humanitarian concern and expanded organizational effort that has taken place in the last fifty years. It means mobilization of the new culture to encourage the wealthy part of the globe, particularly wealthy governments, through the auspices of the UN, but also NGOs, to make the economic sacrifices necessary to create and apply technology in effective ways. As humanitarians have pointed out time and again, the huge challenge of the new century is to redouble our efforts by not only reacting effectively to dramatic crises but also continuing to address the fundamental causes of poverty and lack of infrastructure and economic development.

Finally, this chapter makes clear that technology is, truly, double edged. The same technology that can provide for rapid global reach can enable infectious disease to spread throughout the world. The same technology that can enable coordinated humanitarian actions can also provide devastating weapons accuracy. The same technology that can allow for effective tagging and identification of people can be used to subvert personal freedoms. We are entering a time where we must learn to deal with the dual nature of technology.

In addition, members of the NGO community must accept the

fact that they will need to pool resources in order to provide for the development and integration of technologies necessary for their work. Humanitarian actions are not always considered sufficiently in either military or commercial operations—they often require custom solutions to their own unique problems. These solutions are expensive. Only by coordinating their requirements and resources will the NGOs be able to develop and learn how to use the technologies that can make a real difference in their most important efforts.

COMMUNICATIONS TECHNOLOGIES

The outflow of refugees immediately following the genocide in Rwanda changed the scale of response demanded of the aid community. Aid agencies, previously experienced in coping with establishing camps the size of small towns with populations between 40,000 to 60,000, found themselves confronted with the sudden arrival of numbers equivalent to those of major cities. Tanzania received 800,000, Zaire close to 1 million.

In refugee camps it is essential to register new arrivals. This is done most importantly to give them a legal status, to give them the basis for international protection. There is also the more mundane need to establish who and how many are in the camp for all the usual administrative reasons, not the least of which is to be able to provide appropriate assistance in the right quantity.

Refugees are emotionally at their most vulnerable as they arrive at a camp. They have escaped from the hell that has driven them out of their homes, they have survived the journey, and they now see the first glimpse of protection and assistance. Those who are running the camp will themselves have gone through dramatic times. An influx of refugees is usually overwhelming and very rarely orderly. There is the need to register but there is also the parallel need to offer immediate humanitarian assistance: shelter, water, medicine, and food.

The victims of displacement want to be registered and given a means of identity. They want authorization to be there, to receive shelter, and to stand in the food queue. The challenge to the aid agencies is, amidst the chaos of the crisis, to register the population, to provide emergency services and needs, and to manage the scarce resources.

When the time comes to leave the camps, returnees head for home and often qualify for a reintegration package of food and material assistance. After a time of great hardship and controversy, most of the Rwandan caseload of refugees returned home. In November 1996, over a five-day period, 500,000 refugees living in the camps around Goma in Zaire repatriated to Rwanda. The United Nations High Commissioner for Refugees (UNHCR) estimated that 12,000 returnees an hour were arriving at Gisenyi in Rwanda. Each returnee is also registered.

At times the methods of identification have been crude. Gentian violet, which stains the skin, was used in the treatment of body sores. In discotheques, to permit reentrance, often the clientele have the back of the hand stamped with a dye similar to gentian violet. Not surprisingly it is being replaced by less obvious creams.

In the not-so-distant past in some camps, overwhelmed by arrivals and bereft of resources, one solution for a rapid and temporary registration was to daub the arm of the beneficiary with gentian violet. It was done to the mother of the household as she stepped forward to receive her initial entitlement of food. It prevented the recipient from reentering the queue and receiving a second, unauthorized, issue of food.

It was a solution but an affront to the dignity of the woman and an insult to beneficiaries.

In a refugee camp in the West Bank, where many homes had been destroyed by an Israeli Defence Force incursion, donor funding was found to repair damaged homes. In a number of instances the recipient of the repair grant was the woman of the household. Each grant required a signature or a thumbprint. Those older women not able to write had to roll their index finger over an ink pad. The ink lasts on the fingers for hours, staining clothing and indicating to all that the woman is illiterate.

Not much dignity there either.

Recently in Afghanistan iris recognition was introduced as a means of identification. All the eyes registered sparkled with the novelty of the new technology. No loss of dignity there.

—Larry Hollingworth

Biometrics: Personal ID/Tagging

C. Kumar N. Patel, Ph.D.

1. INTRODUCTION

Halt! Who goes there? Friend or Foe?
Friend.
You may pass.

This is a familiar and time-honored way of identifying people, simple and to the point. What is right and what is wrong with this simple way of identification and authentication? This technique is almost 100% reliable in providing no false negatives.[1] In other words, the probability of false negatives (PFN) is very small. However, the flip side of the coin is that this simple system fails miserably in providing an acceptable level of false positives.[2] Thus, the probability of false positives (PFP) also needs to be small. Different situations will demand different levels of probability of false negatives and probability of false positives.

1. Identifying and authenticating individuals

For example, for security purposes involving highly guarded facilities, some false negatives could be acceptable if alternate or repeat checking is in place. A nuclear facility would fall in this category because the majority of the individuals who would be accessing these facilities would be employees or those who have bona fide business reasons to be admitted. Such a facility needs to have a very low probability of false positives for obvious reasons of safety, to guard against terrorist or otherwise unauthorized access. On the other hand, for access to large public facilities such as public and private office buildings, too high a probability of false negatives would slow down access and cause a significant public problem, as seen often at airports during times of high alert. Here one would like to have a low probability of false negatives, but that will have to be balanced against a high probability of false positives rate that may cause problems.

2. Detecting and identifying unauthorized substances and objects

Consider a situation involving the detection of unauthorized substances and objects as opposed to identifying/authenticating an individual's identity. Prime examples are the screening procedures at airports or detection of chemical or biological weapon agents in the environment. Here the price of false negatives at an airport screening station could be very high, resulting in substantial economic damage as well as loss of life if a miscreant is able to carry a weapon or explosive into the airport and perhaps onboard an airplane.[3] Thus probability of false negatives needs to be exceedingly low. However, probability of false positives also needs to be low to avoid chaos and disruption of traffic flow.[4] It is not uncommon to see some sensing instrument, when it has a high incidence of probability of false positives, turned off, thereby assuring absolutely zero probability of false positives! A prime example of such a sensor is a fire alarm in an office building that goes off very often and results in unnecessary evacuation of the building.

In the present chapter, I will focus on the first of these two situations, the identification and authentication of individuals for a variety of potential applications.

2. ENABLING TECHNOLOGIES

Much of what I plan to examine has been enabled by technologies that have been developed in the last fifty years. As we shall see, while many ideas for personal tagging technologies are new, most of them are quite old, but the communications, computing, and digital storage technologies have enabled them. One key feature underlying both affordability and extensive deployment of the identification technologies is that they rely very heavily on these same developments.

Communications technologies, which saw a digital revolution more than fifty years ago, have benefited enormously from the optical and satellite transmission of information. Much of the long-distance transmission of data takes place via terrestrial and undersea optical fibers, which have replaced copper wires. For many point-to-multipoint types of applications, satellite transmission has proven to be unique. With both of these in place and expanding quite rapidly, the transmission capacity worldwide is growing exponentially. This allows for personal identification data to be disseminated worldwide as needed in a very short period of time.

Computing technologies are advancing very rapidly, making it possible to store large amounts of data, access them, mine them, and use fast computations to compare them with newly acquired data. The speed of computation and the density of integrated circuits have doubled every eighteen months with virtually no end seen in the near future.

3. Values of Personal ID/Tagging

Personal tagging and foolproof identification of individuals can be very valuable for a number of reasons, including:

- Individual's protection: It permits proper identification and availability of legal protection in case of conflict and crisis.
- Protection of a group: An individual can be appropriately approved for participation in the group's activities, assuring that the group is not being infiltrated by an adversarial organization.
- Authentication of information deliverer: Is the individual who delivers the information or an object the person that the individual claims to be? In times of conflict and crisis as well as in normal times, it is important to verify the identity of the information deliverer or the courier. Information delivered by an unauthorized individual should be suspect.
- Authentication of the information receiver: This is the counterpart of the above to assure that information does not fall into the wrong hands, leading to conflicts and fraudulent transactions.
- Security of electronic commerce: As banking and other financial institutions become increasingly connected electronically, large amounts of cash flow over the Internet. Proper identification and authentication of machines is without a doubt one of the

major challenges. A parallel challenge is the unauthorized access to the banking system that results in serious financial loss.

- Information security and database access: With increased centralization and collection of personal information, concerns are rightfully raised about unauthorized access and its consequences to individual privacy.
- Entitlement authorization: This is especially important in a post-crisis situation where attempts are being made to recover from a disaster and there is a need to assure that necessary supplies, food, medicine, and other help reach all who should be getting them.
- Building and public facilities access: With the constant threat of terrorism in public places, we have seen a tightening of security at airports and in large public buildings. However, there are reasons to believe that none of the presently implemented security measures provide the level of confidence that is needed. Furthermore, the price of such security has been a national waste of the population's working time as people wait to get cleared. Thus, personal tagging/ID would go a long way toward greater assurance of security without undue time being wasted for clearance.
- Automobile ignition: A personal tagging system that prevents unauthorized individuals from operating an automobile could go a long way towards reducing the automobile theft crime rate.

4. COMPONENTS AND REQUIREMENTS OF IDENTIFICATION

Following is a partial list of what should be expected of an identification system:

- Identification: Assurance that the person is who she or he claims to be.
- Authentication: Does the person have the necessary credentials for carrying out the tasks that she/he wants to do? For example, in a hospital, does the individual have the necessary credentials for providing medical care?
- Authorization: Does the person have the rights/access that she/he claims?
- Accounting: Keeping track (i.e., record) of individuals who were identified, whose identities were authenticated, and who were duly authorized. There is also a need to keep track of the times when the system did not grant access or authenticate the individual.

There are a number of requirements that any personal tagging/identification scheme must meet:

- The system must be robust against unauthorized disclosure of information that is stored and used in the task of identification.
- The system must guard robustly against tampering from either unauthorized or authorized personnel. The latter is very difficult but is necessary for preventing abuse of the system for private reasons.
- All systems fail. A robust system must have a soft but recognizable failure mode so that other security measures can be implemented in case of a partial failure.
- The system must provide few false negatives. A consequence of high false negatives is that additional time is needed to assure that a nonacceptance of identification is not a limitation of instrumentation. High false negatives at an airport security system would cause chaos because of long lines and the thwarting of legitimate commerce. A system of personal identification that is chronically plagued with high false negatives will eventually be turned off!
- The system must provide few false positives. It is clear that high false positives, i.e., authenticating unauthorized persons, could have very undesirable consequences.

5. Need for Reliable Identification

There are several situations where reliable identification and authentication are very important. These are listed below along with examples and numbers of individuals that need to be screened.

1. Access Control:

Public facilities, such as airports or railroad stations, need to be protected from either casual or determined terrorists who can cause panic and destruction leading to economic disruption. The number of individuals that need to be screened is very large.

Work environments such as offices need to be protected from terrorists as well as industrial espionage activities. The number of individuals to be screened is medium.

High-security environments such as defense facilities or war zones require extremely well-controlled access because the downside risk of a false positive is very high. The number of individuals to be screened is relatively small.

2. *Service Reception:*

Public facilities such as libraries need to assure themselves that patrons are in fact authorized to receive the services they are seeking. This is not a very demanding application since the damage done by a false positive identification is not likely to be significant, and the number of individuals to be screened is relatively small.

Controlled facilities such as hospitals and clinics need to identify and match patients with their prior records as well as their insurance identifications. It is very important that patients and their relevant records are properly matched, but the number of individuals that need to be screened is small, and there is usually ample time for screening.

3. *Service Provision:*

Controlled facilities such as hospitals need to ensure that the individuals providing service, physicians, nurses, and so on, are who they claim they are. The danger to the public from medical care provided by unauthorized personnel can be serious. However, the number of individuals to be screened is small.

Zones of conflict such as rescue missions, relief operations, first responders, and the like, also need to be properly identified and authenticated both for their own safety and for the safety of others. The numbers are small to medium.

6. Types of Identification and Authentication

There are at least four different types of identification processes that one may use, and each of these has various levels of spoofing potential. These are summarized in Table 1 below.

This table clearly suggests that biometrics may provide highly reliable personal identification because it may be tied to something that an individual is born with and may remain invariant throughout a person's life.

Table 1. Types of Identification Processes

Type	Specific	Spoofing Potential
Something one knows	Name Personal information Password/pass phrase	High Relatively high Relatively high
Something one can do	Signature	High
Something one has	ID card Smart card Passport	High Can be made low Relatively high
Something one is	Biometrics	Low to very low

7. BIOMETRICS

Biometrics is a means of using an individual's characteristics that remain unchanged over a long period of time and are unique to that individual. Some of these characteristics and their uniqueness as well as the probability of intentional spoofing or false negatives differ from each other.

1. Facial scan: Fundamentally, this is what we do to identify a person we see. Mentally we are able to process quite rapidly the person's face to make a unique identification. For use as an identification tool, one needs a photograph of the person's face appropriately scanned and digitized for proper analysis. One advantage is that the identification process can be quite rapid. But with the advent of high-technology computer-assisted reconstructive surgery, there may be a high spoofing potential.

2. Voiceprint: Analysis of a predetermined phrase can provide a good tool for distinguishing individuals. However, even with very good analytic algorithms, there is a high probability of false negatives and a high probability of false positives, especially with trained mimics.

3. Signature verification: This has been used extensively because storing a reference template does not require too many bits, but spoofing potential is high. Moreover, a person's signature often changes with time, and therefore the probability of false negatives can also be unacceptable.

4. Fingerprint: This mode of identification has been most often used by law enforcement agencies and now is being proposed for use in identifying individuals coming to the United States from other countries. The FBI has a very large database of fingerprints, and they generally are considered foolproof in most circumstances. Yet, they are capable of being spoofed in situations where fingerprint identification is used by itself in an automated process with no other required forms of identification. It is generally believed that an individual's fingerprint does not change over one's lifetime, and a machine-readable characteristic of a fingerprint does not require a very large data storage (Fig. 1). Because fingerprint identification has been used for more than a hundred years, there is a level of comfort in its usage.

Figure 1.   Example of a Fingerprint Used for Identification

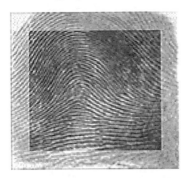

Compared to optical techniques such as iris or retinal scans, fingerprints are left behind by individuals touching most surfaces. Hence, fingerprint identification, which is considered an infallible means of personal identification, is extensively used in criminal and other justice systems through the use of the Automated Fingerprint Identification System (AFIS).

5. Palm scan: This is an extension of fingerprint identification and provides additional levels of information (Fig. 2) with a concomitant increase in the amount of data storage required. The palm scan is a new biometric tagging system with significant potential for unique identification through the establishment of the Automated Palm Identification System (APIS), although there is no large existing database. This is likely to be of greatest use to

Figure 2.   A Palm Print

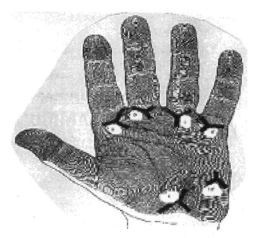

law enforcement agencies and perhaps high-end secure environments.

6. Palm geometry: This identification technique relies on the overall structure of an individual's palm. It holds promise but requires much larger data-storage capability.

7. Hand topology: This technique extends the palm identification technique to the entire hand with expected improvement in false positives.

8. Iris scan: The eye is one internal organ that is visible from outside (see Figs. 3–5). Recent observations of the structure of the iris have shown that a person's iris probably does not change in its fine structure throughout a person's life. It is also believed that an iris scan is virtually spoof-proof and the probability of false positives is negligibly small, especially when utilized in conjunction with other schemes that assure that the individual is not intentionally trying to spoof the recognition system by resorting to extreme measures. Nonetheless, the relatively low intrusiveness of the technique may lead to its extensive usage.

Steps needed for an iris scan personal identification system include:

1. Photograph all the individuals who will have to be authenticated and store their iris "pictures" in a file.

Figure 3.   Externally Visible Eye

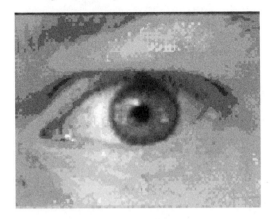

Figure 4.   Details of Externally Visible Eye

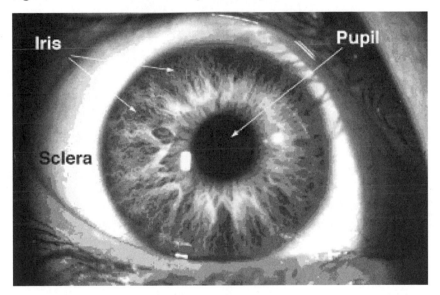

2. Use:

- Acceptable human/machine interface
- Computer vision
- Pattern-recognition algorithms
- Statistics

**Figure 5.   The Details of the Fine and Rich Structure of the Iris
Make It a Good Identification Tool**

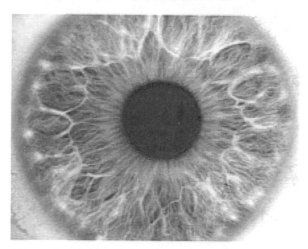

3. Obtain real-time, high-confidence recognition/authentication of the person's identity.

A celebrated case of iris scan for identification is that of Sharbat Gula, a young Afghan girl who was photographed a long time ago. Comparison of the original iris scan (obtained from the photograph) with that from a subsequent photograph of an adult woman permitted unambiguous identification of the two as the same person (Fig. 6).

9. Retina scan: The retina is visible from the outside with some care. This method of identification has been around for a fairly long time and relies on the fact that the structure of blood vessels in the retina (Fig. 7) of an individual does not change from birth to death. The blood vessel structure is unique for each individual, and there is no known way of replicating the retinal structure. Retinal scan for authentication is more intrusive than iris scan. Because of its ultra-high reliability, it is used extensively in high-end, high-security environments even though it is relatively intrusive.

**Figure 6.   Widely Published Photographs of Sharbat Gula as a
Young Girl and a Grown Woman**

**Figure 7.   A Retina Picture Showing the Richness of Blood Vessels
Forming a Foolproof Identification Tool**

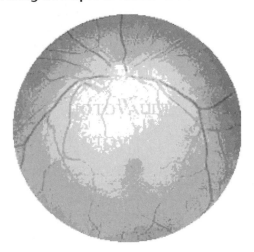

Steps needed to implement a retina scan identification system are:

- Store pictures of individuals' retinal blood vessel structures in a file (ophthalmologists look at the structure to ascertain its health) and create a database.
- Obtain real-time, high-confidence recognition/authentication of the person's identity using sophisticated pattern matching algorithms.

10. DNA scan: An individual's DNA sequence is unique and is the most fundamental biometric signature of human beings. It is genetic and therefore unalterable in sickness, health, or death. It can be very accurate, but for now it is very slow, and there are questions regarding the possibility of obtaining a DNA map without physical contact. Nonetheless, there are good reasons to believe that if rapid DNA scanning can be accomplished, it can be a tool as reliable as the best biometric tool for identification. However, it is also likely that the amount of data storage required could be quite large.

8. Effectiveness, Intrusiveness, and Acceptance of Biometrics

As mentioned earlier, the identification techniques must be effective to be useful. Effectiveness includes both false negatives and false positives. If the system rejects too many bona fide individuals, the identification process will slow down, have substantial economic impact, and the individuals will not tolerate inconvenience.[5] Under these circumstances, the system may even be shut down and not used. If the system has too many false positives, it will not provide the security that is desired.

Another aspect of an identification technique is its level of personal intrusion. A scheme that inconveniences the individual in getting recognized as bona fide is not likely to find widespread usage in situations where alternatives are available. On the other hand, intrusive identification schemes, if they provide a high level of assurance, can be made part of the contract between the individual and an institution, such as employment in a highly classified organization.

In Table 2, I have ranked the various biometric schemes in terms of their effectiveness, intrusiveness, and acceptance.

Table 2.   Various Biometric Identification Techniques Ranked by Their Level of Effectiveness, Intrusiveness, and Acceptance

Effectiveness	Intrusiveness	Acceptance
Retina scan	Retina scan	Iris scan
Iris scan	Hand topology	Signature
Fingerprints	Hand geometry	Voiceprint
Palm scan	Palm scan	Facial scan
Hand geometry	Fingerprints	Fingerprints
Voiceprint	Iris scan	Palm scan
Facial scan	Signature	Hand geometry
Hand topology	Voiceprint	Hand topology
Signature	Facial scan	Retina scan

9. Preparing for Biometrics

Identification and authentication of individuals using biometric tools requires that the data for individuals be stored in some central (or distributed) databanks for comparison when an individual presents herself/himself to be identified. Collecting, cataloging, and distributing the database, whether the information is fingerprints, voiceprints, iris scans, or some other data, requires a substantial amount of preparation, organization, and expense. Table 3 below compares various biometric technologies in terms of the effort needed on the part of the individual and the system to create a large database that will permit widespread use of biometrics for authentication.

10. Biometric Identification System for Humanitarian Actions

The above description makes clear that a perfect system can be devised for personal identification and tagging that has very low

**Table 3.   Effort, Expense, and Organization Required for Creating
Biometric Databases**

Technology	Individual	System
Retina scan	Little	Considerable
Iris scan	Little	Considerable
Palm scan	Some	Considerable
Hand geometry	Some	Considerable
Fingerprints	Some	Considerable
Voiceprint	Some	Considerable
Facial scan	Some	Considerable
Hand topology	Some	Considerable
Signature	None	None

false negatives as well as negligibly small false positives. Generally, the complexity and cost of implementing such a system scales with the size of the population that needs to be in the database and the "perfection" of the system. Depending upon the application, one can design a system that meets the goals.

Case 1. For access to organizations that deal with the national security at the highest level, expense would not be a barrier, and the system must have no false positives. False negatives can be tolerated by having human intervention to provide alternate forms of identification. Furthermore, the time it takes to authenticate needs to be reasonable, but this is not a serious criterion because the number of individuals who need to be in the database is relatively limited.

Case 2. For access to airports, railroad stations, and other public places, the speedy clearing of an individual through the authentication system is crucial. Expense is of moderate importance, but the probability of false negatives has to be balanced against the probability of false positives. High probability of false negatives will bring the system to a grinding halt, and high probability of false positives will make the system ineffective.

Case 3. For humanitarian applications involving authenticating personnel involved in providing services, we need a system that is less tolerant of false positives and provides a moderately low incidence of false negatives. On the other hand, for a system for authenticating the recipients of humanitarian services, it would be desirable to have very low probability of false negatives. If this carries with it a moderate probability of false positives, it perhaps can be tolerated because the consequences of someone receiving food or medical help more often than the person is entitled to are not really serious. However the expense necessary to set up and use the system must be relatively low. Moreover, the preparation and organization needed for implementation cannot be too involved for, in general, humanitarian actions are not as well funded as, say, the military actions that often make the humanitarian actions necessary.

11. CONCLUSION

In conclusion, it is possible to make a perfect system of personal ID and tagging. However, resource limitations will dictate the various compromises that need to be made. In general, in devising a system it is well to keep in mind that a high probability of false negatives results in a socially and economically unacceptable operation. On the other hand, a system that yields a high probability of false positives is dangerous to everyone.

The technology for implementing any of the high-assurance personal ID and tagging systems is here, and if expense is no barrier, a foolproof system can be implemented. However, none of this chapter has touched on the unanticipated consequences of a personal identification system that includes a majority of a national population. These consequences may be sufficiently serious to require that the identification system be compartmentalized to minimize unduly jeopardizing individuals' privacy.

A balanced implementation plan requires that we not forget the adage *Quis Cuodes Custodiet?*

Radio communication is very effective in a relief and humanitarian setting; it is relatively cheap to establish—after the initial purchase of hardware—and costs nothing to run. Radio frequencies are normally allocated to various humanitarian actors by the national government; sometimes channels are shared, but more often than not humanitarian actors operate on different frequencies.

Interventions in camp settings become more complicated due to the vast array of humanitarian actors who are involved. Services provided include camp management; health, water, and sanitation; child protection; food distribution; education; shelter, and the list continues. There can be as many as ten different agencies, both national and international with different capacities and abilities, operating in the same camp. Consistent coordination and communication is essential. Sometimes agencies have shared radio frequencies, but that is not common. In fact, it is extremely rare to have all the agencies in the same camp able to communicate with one another. The importance of communication becomes very apparent during periods of insecurity.

In 2003, I was working in Kandahar in southern Afghanistan. MSF was the health-care provider in a camp for approximately forty-thousand internally displaced Afghans. The camp was literally in the middle of the desert and was called Zhare Dasht ("yellow desert" in Pashtun). Space was not a problem, and the camp was well laid out, but there was a problem with a consistent supply of water and security.

The Zhare Dasht camp was the perfect infiltration environment for Taliban and Al Qaeda who were operating against the Coalition in the war on terror. Tragically in 2003 Osama bin Laden openly declared war against the humanitarian actors in Afghanistan and accused them of working with an illegal government. The assassination of an ICRC delegate was the first in an increasing number of murders of humanitarian aid workers.

By October 2003, security was becoming very difficult. Al Qaeda murdered two humanitarian aid workers who were traveling up the main highway from Kandahar to Herat. Security worsened after an incident occurred in the Zhare Dasht camp. Taliban and Al Qaeda operatives had infiltrated the camp and attacked the demining agents who were working on the periphery removing the mines and unexploded ordinance that litter the country. Thankfully the execution of the aid workers was thwarted, and the operatives fled the scene. That day a security incident was averted, but more ominously only two humanitarian agencies out of seven operating in the camp could communicate with one another to share information. If technology could help provide a fast, user-friendly way of allowing improved communication among different agencies, it could literally mean the difference between life and death.

—Nicola "Nicky" Smith

Wireless Telecommunications

Paul J. Kolodzy, Ph.D.

MANKIND HAS an insatiable appetite for information, and with information comes knowledge. Knowledge helps alleviate distrust, mistrust, and misunderstandings, which often cause conflict in the developing world. Communications is the means to obtain information. The ability to communicate affects all aspects of life, such as interpersonal relationships, financial transactions, and societal trust. This need to communicate can be met through print, voice, or data.

The digital communication revolution in the last decade has provided more technological opportunities to meet the world population's basic need for communications using voice and data, narrowing the so-called digital divide between developed nations and developing ones. The importance of this trend was stated in a World Bank report:[1]

> Access to information and communications technologies has become crucial to a sustainable agenda of economic development and poverty reduction. Communications technologies affect poverty reduction through three primary mechanisms: increasing the efficiency and global competitiveness of the economy as a whole with positive impacts on growth and development; enabling better delivery of public services [such] as health and education; and creating new sources of income and employment for poor populations. . . . Technological innovations, economic pressures and regulatory reforms are making access to information and communications technologies more affordable and providing opportunities to close the digital divide.

The emergence of cellular networks, wireless local area networks (WLAN), and Voice over Internet Protocol (VoIP) are three of the newer technologies that are bridging the communications gap. However, it has been the merging of technologies, both new and established, that has made a much larger impact in making telecommunications accessible in the developing world.

This chapter provides a snapshot of the impact of technological innovation within the telecommunications sector in providing basic communications throughout the world. It will examine the following: Why is wireless technology a key enabler for providing basic communications? What are the choices of technology that can be used as building blocks for the third world? How is the landscape changing for wireless solutions of the future? What are the lessons to be learned from other first users of these new technologies? And what are some of the future challenges and technologies?

1. The Digital Divide

The availability of digital communications throughout the developing world is a growing concern across the international community. For example, the Maitland Report of 1984 raised concerns about the availability of telecommunications technology in rural and underdeveloped areas of the world and the consequences of a world of technological haves and have nots.[2] An International Telecommunications Union (ITU) Report on Internet for a Mobile Generation (2002) indicated a strong relationship between connectivity and GDP per capita (Fig. 1). The high

Figure 1. Relationship between GDP and Telephony Penetration

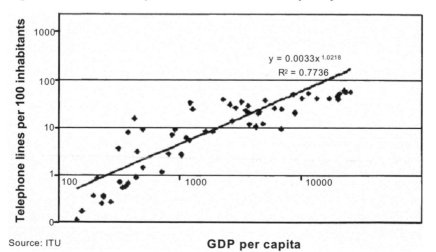

$$y = 0.0033x^{1.0218}$$
$$R^2 = 0.7736$$

Source: ITU   **GDP per capita**

costs associated with Public Switched Telephony Networks (PSTN) often make them unaffordable in many developing countries. This is especially true for rural areas with lower subscriber densities.

In a separate investigation, the ITU's World Telecommunications Development Report concluded that mobile communications could be the key enabler for basic access to telecommunications.[3] Mobile communications can overcome the last-mile connectivity impediment of laying, managing, and maintaining a cable infrastructure. Wireless technologies such as Wi-Fi have the potential to leapfrog technology for countries lacking a wireline infrastructure (ITU Telecom World 2003). China is one example, using this leapfrog technology to address its challenge of reaching rural communities by using wireless access.

Both the Maitland Report of 1985 and the more recent ITU World Telecommunications Development Report highlight the criticality of providing digital connectivity to the remote communities of the world. The difference between 1985 and 2003 has been the availability of new, cost-effective wireless technology.

However, concerns about a region's ability to sustain a viable or profitable customer base have impeded telecom development in certain places even with these cost-effective technologies. Hamadoun Toure, director of the ITU Telecom Development Bureau, challenges that opinion. He has stated: "The myth in the information society is that people in developing countries either can't afford or are unwilling to pay for ICT services." Demand aggregation can address this problem. For example, GrameenPhone in Bangladesh coordinated with local entrepreneurs to build a cellular phone network and customer base. GrameenPhone noted that although individuals may be too poor to be attractive customers for profit-seeking businesses, those individuals collectively represent a valuable customer base. Strategies designed to meet this need should take into account the rapid development of cheap, efficient telecommunications technologies.

Clearly, cost-effective communications using innovative business models that exploit advanced wireless telecommunication technology are feasible, at least in some demonstrated cases. The United Nations, however, is attempting to find the means to expand on the localized success stories to create a more global effort. The UN's Information and Communications Technologies Task Force in 2002 stated: "One of the most pressing challenges [for communications] in the new century is to harness this extraordinary force, spread it throughout the world, and make its benefits accessible and meaningful for all humanity, in particular the poor." The task force is working to identify successful solutions and standards for access networks providing open universal access (first mile), as well as sustainable business models for operation and maintenance of such networks, which could be disseminated as best practices.

First Mile or Last Mile?

Is communication connectivity considered the first mile of the network or the last? The common focus on most telecommunications projects for humanitarian aid has been on extending the World Wide Web, or Internet, to lesser-developed countries—that is, extend the edges of the network. The "last mile" perspective indicates that the "remote" edges are the last steps in the architecture. Cellular systems are a classic example of "last mile" architecture since the cellular towers were an extension of the telephone network at the edges. Is this an appropriate model for the lesser-developed world?

The last mile perspective does not apply well to regions that lack the network infrastructure and thus places the onus on the users to convince the providers to extend the network. Titus Moetsabi challenged this perspective. In 1997, Moetsabi coined the term "first mile" to change the networking architecture paradigm. The focus, per Moetsabi, was to begin with the edges and have the infrastructure within the urban centers be focused on how to interface with those edges. To that end, rural telecommunications technologies need to be designed with rural people as active participants in all phases of the development.[4] This is a far less top-down approach to the challenge of providing universal connectivity, regardless of income or location.

Low-cost technology is available to develop cost-effective rural networks, but lack of awareness of the technology and lack of skill in using it are formidable obstacles to its deployment. So what are the technology choices in which to build first mile/last mile access?

WIRELESS TECHNOLOGY CHOICES

The advent of cost-effective, device-oriented technology has reduced the barriers for deploying wireless connectivity in both developed and developing countries. Technologies for wireless connectivity include wire, microwave, cellular, multi-access radio, and satellite. Many parameters influence the selection of a technology solution. However, two basic parameters have the greatest impact: user population density and the distance from the main network. As shown in Fig. 2, wire and fixed microwave systems are the predominant technology choices in highly populated centers, because the costly infrastructure associated with these systems is amortized across a large population base. However, in areas of low population density, predominant in lesser-developed countries, this is not the case. In these cases, a wider range of technology is necessary to meet the cost and communications requirements. The primary trade-off is time. If voice traffic can be

Figure 2. Telecommunication Technology Utility

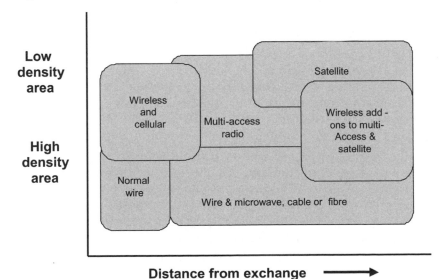

supplanted by E-mail, then significant savings can be obtained. If instantaneous E-mail can be supplanted by periodic E-mail, then additional savings are afforded. The following section describes the variety of technology choices that are currently available to the communication developer.

SATELLITE

Satellite communications (SATCOM) systems provide rapid, accurate communications almost anywhere in the world. Geostationary satellites orbit at about 22,000 miles directly above the equator. At that altitude, the satellite orbits the earth at the same rate as the earth's rotation, so they appear to remain stationary in the sky. This configuration allows an individual satellite to cover a single hemisphere and thus can be very cost-effective in terms of the number of satellites necessary for global coverage. Low earth orbiting (LEO) satellites orbit at much lower altitudes, as low as a few hundred miles. These satellites orbit the earth much faster than the rotation of the earth itself and thus will only stay over a particular area for a short period of time, as little as a few minutes. Therefore, a constellation of multiple spacecraft is needed to provide global coverage. In either case, the high cost of developing and launching a spacecraft and the cost of SATCOM services have been too high for widespread use by the humanitarian action community.

Geostationary satellites have been used for many years to provide voice and data services to mobile terminals. However, due to the long transmission distances, the signal levels are quite weak once they travel the long distances between the satellite and the ground station. One limitation is the amount of power available at the satellite for the downlink transmission. If the ground station is mobile, then the same limitation is at the ground station to produce a sufficiently strong enough signal to be detectable at the satellite. In general, either the ground station must have a large collection dish or the data rate must be reduced in order to have enough signal to close the link. A final limitation is when the signals are so low that an insufficient signal will not penetrate a building, so the ground station antenna must have a direct line of sight to the spacecraft. Therefore, geostationary satellites allow communications anywhere in the world with very few spacecraft. However, limitations in data rate and the requirement for large antennas can make the system impractical for remote locations.

LEO satellite systems are much more suitable for communicating with relatively small, low power, handheld ground stations. However, due to the short duration of coverage for a single satellite, more complex access and relaying techniques must be employed in the satellite. Iridium was an LEO constellation developed for satellite telephony. The switching was accomplished between the sixty-nine satellites, thus increasing the complexity, and thus cost, of the satellites. Globalstar was another LEO constellation developed for telephony. However, Globalstar was a bent pipe that simply relayed any telephone signal to a network access point on the ground. Therefore, a series of ground station access points needed to be constructed.

Very Small Aperture Terminal (VSAT) technology has been developed to provide a cost-effective solution for the SATCOM ground terminal. VSAT systems can use a relatively small (one- to two-meter) antenna. A popular configuration for remote communications system is to use a VSAT SATCOM ground station as a hub that is connected to multiple ground elements in a "STAR" configuration.

MULTI-ACCESS RADIO

Wireless communications use a shared resource, the electromagnetic (aka radio frequency, RF) spectrum. The RF spectrum is a finite resource, and there are not enough frequencies, or channels, to allow each user to have a dedicated channel. Therefore, channels must be shared among users. Older systems employ push-to-talk (PTT) mechanisms. PTT has a new user listen to the voice traffic on a channel and when a lull in the traffic occurs, the new user will then transmit. There are three basic problems with this mechanism:

1. Although a new user may sense a "lull," another user may sense the same lull and attempt to transmit at the same time, causing interference to both;

2. In order to sense a lull, the new user listens to all the other traffic on the channel, potentially creating a privacy and security risk; and
3. The PTT mechanism relies on all the users to follow the etiquette.

The PTT mechanism has been fairly successful and has been used extensively in systems such as those for public safety, private land mobile, and citizen band (CB) radios. However, PTT is a manually executed protocol and thus is limited in how much of the channel capacity can be used due to delays and inadvertent interference.

Advances in technology have provided an additional channel-sharing mechanism that uses the channel capacity more efficiently and eliminates the base problems associated with the PTT mechanism. In multichannel trunked systems, computer logic assigns channels from a pool of frequencies and recovers them at the end of a transmission. When a user wants to send a message, the mobile unit sends out a burst of information, which identifies the individual mobile unit with which the users want to communicate. The computer logic identifies an idle channel in the pool and tells the user and the mobile unit to move to the idle channel. If all the channels are busy, then the call request is placed in a queue and is handled on a "first-in, first-out" (FIFO) basis.

CELLULAR

Cellular wireless services utilize hub-and-spoke architectures to provide mobile connectivity. Wireless cellular telephony has proven to be an overwhelming success by providing access for 1.2 billion users in 2002. This success is even more astounding when one notes that cellular wireless has also eclipsed the fixed wire telephony service in the number of users. The hubs, or cellular towers, are connected to the fixed wire infrastructure and thus cellular addresses the last mile connectivity requirements for voice (and with 2.5G/3G data). However, the necessity of the fixed wire infrastructure has facilitated its implementation only in developed countries where the infrastructure already exists. The utility of cellular infrastructure, as currently architected, is quite limited for the rural areas where there is a severe limitation in infrastructure.

The cellular wireless service has migrated from analog voice transmission (1G or AMPS) to digital voice transmission with circuit-switched data (2G using CDMA, GSM, TDMA protocols) to digital voice and packet-switched data (2.5G using CDMA/ 1xRTT, GSM/GPRS). The next generation, or 3G, cellular systems began being deployed in 2002 in Asia and in 2003 in Europe and the United States. These services include a mobile data-on-demand capability that will range from 384 kbps up to 2 mbps, depending upon the mobility of the user.

The cellular wireless data services, the core feature of the 3G networks, have been challenged by the equally successful wireless local area network (WLAN) technology. Service-based networks, such as cellular, focus on guaranteed service and thus require significant infrastructure. Device-based networks, such as WLANs, focus on ease-of-access with consumer-level devices and require less infrastructure but do not have guarantees of service quality or adequacy.

WIRELESS LOCAL AREA NETWORKS (WLAN)

The U.S. Federal Communications Commission determined in 1985 that the robust operating characteristics of spread spectrum waveforms could allow a new type of spectrum access: unlicensed or license-free. These characteristics afforded operation in normally "trash" spectral bands such as the Industrial, Scientific, and Medical device bands (900 MHz, 2.4 GHz, 5.8 GHz). In reality, this was device-level access. The devices were "licensed by rule," requiring the device to conform to the transmission limits set by the FCC but without protection from interference by other unlicensed devices.

In the early 1990s, initial devices were available such as the AT& T WaveLAN card operating at 902–928 MHz. By 1997, the IEEE 802.11 working group developed standards for the 2.4 GHz ISM band (2400–2483 MHz), allowing three channels of 23 MHz that eventually became the 802.11b standard, providing data rates of up to 11 mbps. In addition, the 802.11a standard was developed for the 5.8 GHz band, providing data rates up to 54 mbps. Recently the 802.11g standard was released that combines the three channels at 2.4 GHz and utilizes a more efficient OFDM waveform to provide up to 54 mbps data throughput.

The IEEE 802.16 standard addresses the first mile/last mile connection in wireless metro-area networks (MANs).[5] The primary focus of the standard is on Broadband Wireless Access (BWA). It operates between 10 and 66 GHz with extensions to 2 to 11 GHz with optional peer-to-peer and mesh topologies. It defines a medium access and control (MAC) layer that supports multiple physical layer specifications tailored to the frequency band of use. Bandwidth will be greater than 20 MHz, providing data rates in excess of 30 mbps and up to 134 mbps for fixed topologies. Products were to be shipped by the end of 2004.

The IEEE 802.20 is under development for Mobile Broadband Wireless Access (MBWA) to enable worldwide deployment of cost-effective, spectrum-efficient, always on, and interoperable mobile broadband wireless access systems. It will provide symmetric, guaranteed latency data services to the fully mobile end-user. The operation is to be in licensed bands below 3.5 GHz. The bandwidth is to be less than 5 MHz with a spectral efficiency of greater than 1 bps/Hz, utilizing advanced technologies such as smart antennas and OFDM modulation.[6]

INFRASTRUCTURE AND AD HOC NETWORKING

The selection of radio technology provides the point-to-point links that a network will use to provide connectivity for a region. There are two primary types of networks that can be used for wireless systems: infrastructure network and ad hoc networks.

Infrastructure networks have access points with which clients' radios will communicate (Fig. 3). The access points are connected to wire or wireless backbones, which are then connected to other access points. Client radios can only communicate with each other by relaying through the access point. Therefore, the mobile client radios can be less sophisticated since they can connect only to one other radio, the access point, at a time. This is how most cellular voice and wireless data systems are configured.

Figure 3. Infrastructure Versus Ad Hoc Based Networking Provides Different Choices for Mobile Users and New Concepts for Extending the Operational Range of the Network. Complexity of the Radios Increases since Each Radio Must be Able to Operate as a Base Station

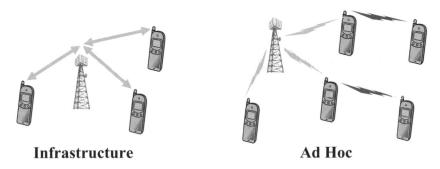

Infrastructure   **Ad Hoc**

However, with the advent of more digital signal processing and programmable analog devices, a new type of network has evolved called ad hoc networking. Ad hoc networking pushes the capability of the base station down to each of the client radios. In that way, each radio becomes a base station (Fig. 3). In the infrastructure mode, a client radio must be sufficiently close to the base station in order to communicate and thus to be part of the network. In ad hoc networks, the client radio only needs to be close enough to another client radio to be part of the network. Then a message from one client radio is "relayed" through many other client radios until it finally reaches its destination. This type of network allows for networks to be formed quickly but requires much more sophisticated radio and routing software within each radio. This technology has shown promise in military network research and is a very active research and development area in wireless networking.

Radio Choice Summary

Overall, the developers of wireless networks for humanitarian actions have a rich diversity of technologies from which to choose. The choices of radio parameters depend upon the operational parameters, such as the distance to the exchange or Internet

Table 1. Sample of UHF, Microwave, and Millimeter-wave Radio Parameters

Communicator	Frq.	Power	Range
Cell phone	UHF	260mW	10km
Adv. Digital Radio	UHF	5W	32km
Portable comm-station	UHF	270W	160km
Vehicular comm-station	UHF	2kW	320km
Fixed comm-station	UHF	4kW	320km

Civilian communicators

Communicator	Frq.	Power	Range
SatPhone	Microwave (C-Band)	1W	300km
Portable Satlink	MMW	35W	7,500km
Vehicular Satlink	MMW	100W	25,000km
Fixed Satlink	MMW	150W	50,000km

Satellite communicators

Type	Frq.	B/U	P/B	Range
LEO Commercial Satellite	MicroWave (C-Band)	20/5	2.5W	250km
LEO Military Satellite	MicroWave (Ka-Band)	40/10	10W	500km
GEO Commerical Satellite	MMW	20/30	100W	1000km

Communications Satellites

B/U denotes (number of beams/number of users per beam).
P/B denotes transmission power per beam. Beam Diam. Denotes beam diameter.
Data/U denotes the available data rate per user.

backbone and the available power. Table 1 provides a sampling of available systems in the UHF, microwave, and millimeter wave bands of the spectrum. As shown in the table, long-range communications via ground-based systems require a great deal of transmitted power (2,000 watts for 320 km). Such power requirements limit the areas of use to where commercial power is available. The satellite communications systems require much less power but require the space infrastructure.

NEW PARADIGMS FOR CONNECTIVITY

The unique challenges in providing data communications technology to remote and lesser-developed regions of the world have provided a breeding ground for unique solutions. Lack of infrastructure (power, connectivity, electronic equipment) is a primary obstacle to overcome. Low user densities limit the amount of return on investment, either in economic or humanitarian terms, that could be used to overcome the lack of infrastructure. Therefore, solutions requiring little or no expenditures on new infrastructure are ideal. Backhaul connections are key components that need to be addressed.[7]

Figure 4.   DakNet Store and Forward Wireless

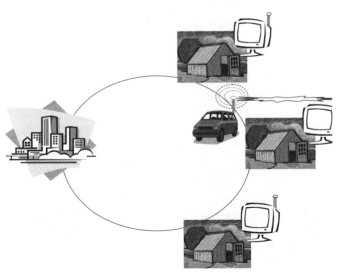

New methods that exploit already existing infrastructure and/ or reduce the requirement for the communications system have been developed. For example, DakNet, which is sometimes called "the Wi-Fi postal system," is a patented wireless network connecting remote regions with each other and the national network. The premise of DakNet is that these remote communications can be tolerant to delay. E-mail was chosen as the ideal application. Although E-mail does not provide interactions in a real-time manner, it can still provide a quantum improvement in connectivity to remote, low-density areas of the world.

Figure 5.   DakNet Kiosk

DakNet was conceived, developed, and patented at Massachusetts Institute of Technology (MIT). It is an example of a first mile solution described earlier. The government of India provided seed funding to the MIT-established Media Lab Asia in 2001 to construct the DakNet project under the Bits for All Initiative program.

DakNet uses "store and forward" techniques to provide long-range connectivity. Locally, kiosks are constructed with personal computers and E-mail software. Generally there is one kiosk per town or village. Each computer also is configured with a wireless 802.11b Wi-Fi Wireless Local Area Network (WLAN). The kiosk has no direct external connectivity. E-mail messages are written and stored on the local computer. Periodically a connection is available to the external network, and the messages are then forwarded to and from the E-mail server. This allows for a global E-mail capability without a persistent connection.

The external network connection technique used by DakNet is extremely clever. A van, or in some instances a motorcycle, is configured as a mobile E-mail server complete with a laptop computer and a wireless LAN card. As the mobile E-mail server comes within range of the kiosk, new E-mail is uploaded from the kiosk and downloaded from the van. Due to the untethered nature of wireless, the mobile server does not have to be very close to the kiosk and does not actually have to stop. After the E-mail exchange, the van then moves on to the next village and its kiosk. The van continues from village to village and then eventually to the larger town or city where Internet access is available and uploads and downloads any messages (Fig. 4).

Case Studies

The International Telecommunications Union (ITU) monitors the modern-day issues in telecommunications and how individual countries have addressed them. The following case studies provide some insight as to how various countries are meeting their telecommunications needs through the use of advanced communications technology. Many of the case studies are exploiting the cost-effective and ease-of-use characteristics of wireless LANs. The advancement of wireless LAN technology has had a significant impact on how remote, lesser-developed parts of the world bridge the communications technology gap.

Latvia and Moldova. The Eastern European countries of Latvia and Moldova are developing countries with poor-quality telephone infrastructure.[8] Because of the lack of such infrastructure, high-speed wireless Internet links have provided a cost-effective and fairly rapid deployment option for interconnectivity.

Figure 6.   OAUNET Configuration with VSAT Long Haul Link and WaveLAN 900 MHz WLAN

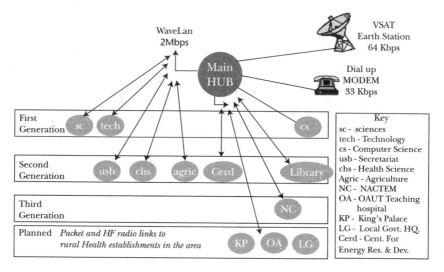

The University of Latvia installed the first citywide wireless LAN link (LATNET) in the capital of Riga in 1993. The genesis of the idea to install such a system came from a Cylink demonstration during INET '93.[9] This first-generation system used the AT&T WaveLAN (now Agere Systems) equipment operating in the 902–928 MHz band. The 2.4 GHz band was not select at that time due to equipment cost. The second-generation system was started in 1995 and operated on the Aironet Arlan (now Cisco) equipment operating in the 2.4–2.485 GHz ISM band. In 1997, a second GSM mobile telephone operator was assigned the band, and the 902–928 MHz band was abandoned.

The configuration as of 1999 had more than two hundred sites in Riga and in towns across Latvia. In many locations the architecture is a single, higher elevation, omni-directional antenna on a rooftop providing access for up to 10 km. For higher usage areas, such as Riga, multiple "feeder" sites were installed with high-speed trunk lines. Arlan bridges were deliberately rejected for this project because they reduced the available bandwidth for data to the end user. Internet connectivity was provided through two satellite links: a 3 mbps link through Crawford Communications and a 2 mbps through Taide Network.

The planning for the first Internet link in Moldova (moldova.net) began in 1996. The poor local telephone infrastructure motivated the development of a wireless network modeled after the LATNET system in Latvia. The initial configuration was a wireless bridge with an Ethernet router using 2.4 GHz Aironet equipment with directional antennas (24 dBi) providing for communications ranges up to 40 km per link. Eventually these bridges were converted to wireless routers. The wireless bridges were then used to further extend the wireless LAN coverage to more remote locations around the capital, Kishinev.

Nigeria. In 1995 the Abdus Salam International Centre for Theoretical Physics in Trieste, Italy, initiated the Programme of Training and System Development on Networking and Radio Communications.[10] The objective of the program was to bridge the technology awareness gap mentioned earlier by providing technical assistance and training to academic and scientific institutions in developing countries. A collaborative pilot project between Abdus Salam ICTP and Obafemi Awolowo University (OAU) of Ile-Ife in Nigeria helped develop OAUNET. The campus network has been in operation since 1996. In the summer of 1997, the project extended the network by helping to set up a radio link between the two campuses of Bayero University within Kano, Nigeria.

The wireless network depicted in fig. 6 shows a Very Small Aperture Terminal (VSAT) providing Internet access with a WaveLAN operating in the 902–928 MHz band. This is quite similar to the configuration used in Latvia in the mid-1990s.

Solomon Islands. In the Solomon Islands, communications between trusted family members and professional peers have been problematic. Sociologists believe that the lack of trusted information, misinformation, and the absence of communications have only heightened the misunderstandings and tensions between communities.[11] These misunderstandings are most acute in deprived and remote areas where telecommunications are least available. Outside the capital of Honiara, telecommunications are very scarce. Because the islands and the population are so scattered, fixed wired infrastructure is not a cost-effective solution. The tele-density in 2002 was 0.3 per 100 rural inhabitants.

Table 2.   Sample Utilization Profile from PFnet from 4547 E-mails (source http://www.peoplefirst.net.sb)

Family correspondence	2038
Business and investment	442
Education	409
Project / NGO	250
Ordering Supplies / Stocks	249
Health / Medical	235
Travel	201
Church	167
Finance and Banking	157
Other/Unknown	49
Construction	49
School fee	43
Sports	30
Government Administration	18
Lands and Titles	17
Agriculture	9
Fisheries	6
Women issue	5
Forestry	4
Emergency	2
Police/Law and order	2

Under the United Nations Development Program (UNDP), a Pfnet system was developed in order to offer a basic wireless E-mail system. The goal of the system was to improve connectivity while significantly reducing the price of communication. The technology enablers for Pfnet were recent advances in shortwave (HF) radio and solar power technologies. Pfnet selected the Pactor Wavemail system of Schuemperlin AG, Switzerland, that incorporates the Swiss PTC-II HF modem.[12] The PTC-II modem provides 589 bit/sec uncompressed data, equivalent to 1,000 bit/sec plain text using the automatic data compression transmitting. The HF radios used included the ICOM IC78, which can transmit up to 100 W. An 80 W solar power supply was used to power the remote stations, which are readily available in the marketplace.[13] On a periodic schedule, several times a day a remote E-mail station will connect via the HF radio to the station in Honiara automatically. During the connection, incoming and outgoing E-mails are transferred, if necessary, to the Internet.

The measured effect of the Pfnet project, as of early 2003, has been a sustained E-mail usage of hundreds of E-mails per month per station (Table 2). As of April 2003, nine remote stations were installed with plans for an additional eight in the near future.

Bhutan. Bhutan is located in the northeastern Himalayas with a population of 600,000 and a tele-density of 3%.[14] Bhutan Telecom, a 100% state-owned corporation, was responsible for implementing the Information and Communication Technologies (ICT) Master Plan of Bhutan. Three main factors were considered in the plan:

- Increase the penetration ratio
- Set human resource development as a high priority
- Explore cost-effective and reliable technologies.

A pilot project was commissioned in 2002 to connect schools and community centers in eighteen villages.

The 2.4 GHz ISM band was the selected frequency due to the availability of equipment and the better propagation characteristics. Two locations were developed: one in Limukha and the other in Kelephu. Each location was a star configuration with the hub connected to the Network Operations Center in Thimphu. In Thimphu the VoIP subsystem was connected to the PSTN. The Voice over IP (VoIP) component of the system was supplied by VocalTec.[15] Each location had repeaters that were up to 10.5 km from the central hub. A conservative estimate would be about 12–15 km between 8dBi omni antennas and about 25–30 km between 24dBi dish antennas using one-watt amps. The range requirements coupled with serious power deficiencies at the remote sites required trade-offs to be made between increasing power and antenna gain.

The availability of power is a serious constraint for wireless tele-communications. The power constraint came in two forms in Bhutan: with commercially available power and with solar power where commercially available power did not exist. In the case of commercially available power, batteries along with a charger were required at the remote site. For sites without commercial power, 70-watt solar panels were used in addition to the batteries and charger. A great deal of care is necessary to compute both the power requirements and the means to supply power. Solar loading and characteristics of individual solar panels can vary widely and adversely impact operations.

Specific systems design parameters included:

- Security, due to the number of communications elements to which the customers have physical access
- Authentication for each device in the VoIP network
- Privacy, since the wireless traffic can be easily monitored with software readily available on the Internet
- Call data records should be in a form easily usable by many third party billing systems.
- Call flow monitoring to collect useful statistics on system usage and state

Applications for the pilot project included telemedicine, tele-center kiosks, and E-post offices. The initial implementation for telemedicine was a connection between two hospitals through a 64 kbps leased line. The tele-center kiosks provided Internet, telephone, facsimile, and E-mail services at reduced costs. The first kiosk was located in Jakar. The E-post offices were constructed using a community E-mail address from which each E-letter would then be printed out and delivered to the concerned home within the locality.

Tomas Evslin, CEO of ITXC, which is purported to be one of the largest VoIP carriers in the world, said of the Bhutan system: "It is an indicator of what the future will look like."[16]

REGULATORY CHALLENGES

The radio frequency spectrum is the medium in which wireless communications take place. It is the natural resource without which the communications systems described earlier in this chapter would not exist, and it is a finite resource that can be accessed and reused in many ways. Access to the RF spectrum is allowed by the regulatory agency within an individual nation. The International Telecommunications Union (ITU) provides a coordination role around the world in order to facilitate interoperability of equipment and international boundary issues. The International Engineering Task Force (IETF) provides a coordination role for standards in many areas, including telecommunications. But the individual nation determines who has access to RF spectrum resources and for what purpose.

One of the primary roles of the Federal Communications Commission in the United States is to coordinate the various uses of the spectrum to prevent different systems from interfering with one another. This is accomplished through the licensing process. Prior to the advent of low-cost electronics, only large entities could afford to build communications systems.[17] However, in the mid-1980s, the FCC determined that low-cost, low-power communications systems were technically feasible and that a new class of license might be necessary to accommodate these devices. The FCC determined that it would be infeasible to provide each individual a separate license. Therefore, a new class of license, within the Part 15 Rules of the FCC, was developed. This license was a "license by rule," which meant that anyone could own a device, but the devices had to conform to particular radio parameters. The limitation was that the devices would not have any interference protection from the FCC from other devices. Multiple bands were designated where these devices could operate (for instance, 900 MHz, 2400 MHz, and 5800 MHz). These bands and subsequently the "licensed by rule" devices were called the unlicensed bands and unlicensed devices.

However, the unlicensed bands and devices were a development within the United States. In many countries around the world, the regulatory agencies have not assigned these unlicensed bands due to political and administrative reasons. "We see many of the actions of the developing countries to be scarcity mentality, meaning, to minimize the amount of unlicensed spectrum to maximize their regulatory position for a few dollars of regulations or licensing benefits."[18]

One example was in Mozambique, where the South African ISP (Internet service provider) UniNet Communications faced regulatory obstacles in setting up a wireless network. The developers complained that the government was initially "very obstructionist," and UniNet had to lobby persistently for the permissions it needed.

A second example was in Nigeria with NITEL, the state-owned telecommunications company. NITEL was the only company that handled telecommunications within and outside the country. Other companies were not allowed licenses. It took international pressure, especially from the World Bank and the International Monetary Fund, to allow a government-backed company to join NITEL to provide international links. This action provided OAUNET the gateway to the Internet through a 64 kbps VSAT. It must be noted that if not for the regulations, especially on the international link, OAU would have been connected much sooner through various international collaborative projects.

Therefore, the impact to humanitarian projects could be significant. Although the technology is available and cost-effective, it may not be allowed due to national regulatory impediments, and significant planning and persistence may be required to overcome those impediments.

Some problems also arise within the context of interference, but generally simple engineering solutions can be provided. For example, in the Bhutan project, the microwave towers that some of the repeaters were on had a preexisting service known as Digital Radio Multiple Access Subscriber System (DRMASS), which was also in the 2.4 GHz band. The channel used by DRMASS was below the ones available to the 802.11b radios but close enough that mutual interference was an issue. To resolve this, the backbone dishes were changed from vertical polarization to horizontal polarization, and then different channels were tried.

Conclusion

The use of wireless telecommunications technology for humanitarian actions will continue to change, for the better. The changes are occurring in both long-range connectivity with the onset of more cost-effective satellite services and in localities with the onset of mobile data networks. The primary choices for agencies embarking on a new mission must take into consideration the need for real-time connectivity and the trade-offs between local, regional, and global connectivity. The choice between voice and data has been essentially eliminated, since voice communications are capable via data networks using VoIP technology. When the requirement for real-time connectivity is relaxed, the number of choices increases significantly. The connectivity range determines the type of technology that is needed, and costs increase proportionally with transmission distance.

Lesser-developed countries have neither the financial resources nor the population density to build infrastructure for communications. Therefore, communications must come from either reduced infrastructure or infrastructure-free techniques. Mobile wireless communications provide cost-effective choices to less advantaged countries. The reduction in the cost of the infrastructure, or more importantly, the development of infrastructure-free systems has made the most impact. The data networking community is developing the infrastructure-free technology for mobile applications. Although the end-user requirements are quite different from those for humanitarian actions, the power and connectivity requirements are very similar. Therefore, the technologies are well matched and a source for continued improvement.

Local and regional connectivity are where the major changes are occurring. Due to the commercial success, the price point for devices has dropped considerably. Coupling these devices with advanced antennas can provide regional connectivity. Although individual technology components for communications continue to improve, the use of hybrid systems is increasing at the same time. Ingenious use of low-tech and high-tech solutions is providing new ways to improve the communications networks in remote areas.

Developing the means to provide communications is only one-third of the solution. Improving the cost-effectiveness of those systems and the ability to deploy them make up the remainder of the solution. An individual nation's regulatory community can provide obstacles in the deployment of advanced communication systems. The obstacles can be based on financial, political, or administrative reasons.

Those involved with humanitarian actions should be enthusiastic about the new tools available for communications. Many of the data network technologies can be easily adapted for regional communication needs. But the future, hard enough to predict with any technology, is even harder with wireless technology due to the pace of innovation. The humanitarian community should frequently monitor two communications technology areas: power generation and directional antennas. The reliance on power is a primary limitation to all wireless communications systems. The current state of the art of 100 W solar systems will be a limiting factor for many deployments. Additionally, directional antennas provide greater coverage given the same power, and power has been, and will always be, a limiting resource in isolated areas. Advances in directional antennas, motivated by urban data networking needs, will provide new capability to extend communications from the local, to the regional, to interregional communities. The humanitarian aid community should periodically review advances in cost-effective solar systems, the requisite battery backup systems, and directional antennas for unlicensed devices.

More and more the humanitarian sector operates in areas where armed aggression is common and violence a constant threat. Regions, villages, roads, and checkpoints are categorized as go-or-no-go areas, threat levels rise and fall, sometimes by the day, sometimes by the hour. The knowledge of present or imminent danger is one-half of a lifesaving equation; the other is the dissemination of the information that can clear the streets, reroute vehicles, halt convoys, and make both the aggressor and the innocent aware of each other's presence.

There is neither a UN agency nor a non-governmental organization unable to tell a story of communications chaos within its own family, where one group has been unable to contact another group with consequences ranging from the simple and not too important to the deadly and devastating. Similarly, agencies and organizations can tell tales of confusion and catastrophe that could have been avoided if they had been able to talk to each other's organization.

Working in war zones fine-tunes and challenges the principles of neutrality and impartiality and independence, especially where aid bodies are working regionally or cross- border and need to negotiate with all sides of the conflict. Who to deal with may be difficult to discover; how to contact them may be next to impossible. Contacting one side may compromise relations with the other.

Humanitarian pauses and peace negotiations are long and tedious processes demanding contact with all sides, and once reached, agreements need constant monitoring and an instant inbuilt system for reporting breaches if they are to hold. Open and rapid communications are the key to success.

Sadly in war there is always the possibility of "blue on blue" or "own goal" attacks where neutral, innocent groups are attacked usually from the skies by "friendly" forces whom they can clearly see but cannot contact. The consequences of such an attack have been graphically summarized by BBC cameraman Fred Scott, who was close to a Kurdish militia and U.S. special forces patrol caught in an American missile attack outside Kirkuk in April 2004, which cost a number of lives including Kamaran Abdurazaq Muhamed, one of the teams' fixers.

"For me everything about that missile strike defines what I've witnessed and filmed in war. It was an instantaneous, unaccountable, intimate, terrifying waste of life and promise at a nameless junction in the hills. After it had rained twice you would never have known what happened there."

—Larry Hollingworth

Cognitive Radio for Humanitarian Operations

Joseph Mitola III, Ph.D.

THE DEGREE OF SUCCESS OF humanitarian operations often depends on the ability of the relief organizations to communicate with each other, with indigenous relief efforts, with distant support organizations, and with those distressed. Voice communications are crucial, but digital communications play an increasingly important role. Emerging wireless technology offers the promise of both better communications connectivity and new technology-enabled relief tasks such as digital triage and telemedicine. Such applications, however, require significant radio frequency agility and bandwidth, some of which will be provided by third-generation (3G) and 4G technologies.

But legacy radio technologies must deliver much of the required connectivity because few can afford to acquire the new technologies instantaneously. Even with substantial indigenous wireless infrastructure, the radio spectrum in a given relief area rapidly becomes crowded by a concentration of aid organizations in bottleneck locations such as cities, ports, and stricken areas. Increasingly intelligent radios are emerging: from today's software-defined radio (SDR) through various modes of aware-adaptive radios to the ultimate cognitive radio.

They offer mechanisms for the flexible use of radio spectrum and the flexible delivery of communications services. To manage the flexibility, radios must employ awareness sensors and adaptation protocols that may be termed radio etiquettes. Aware-adaptive (AA) radio is a particular extension of software radio that employs model-based reasoning about the radio itself, about users, multimedia content, and communications context. Cognitive radios (CR) are AA radios that learn from experience. This chapter characterizes the potential contributions of SDR, AA, and

Approved for Public Release by DARPA [44312] on 18 Nov 03 Distribution Unlimited

CR to improve communications services and spectrum efficiency, highlighting the potential benefits to humanitarian operations.

1. BACKGROUND

A. *Software Radio and SDR*

A software radio is a multiband radio capable of supporting multiple air interfaces and protocols through the use of wideband antennas, RF (radio frequency) conversion, analog-to-digital Converters (ADCs), and DACs.[1] In an ideal software radio, all the aspects of the radio (including the physical parameters of the antenna and multiple physical-layer air interfaces) are defined in software on general-purpose processors. For some air interfaces such as Wideband Code Division Multiple Access (W-CDMA),

Figure 1. Digital, SDR, and Ideal Software Radio Characteristics

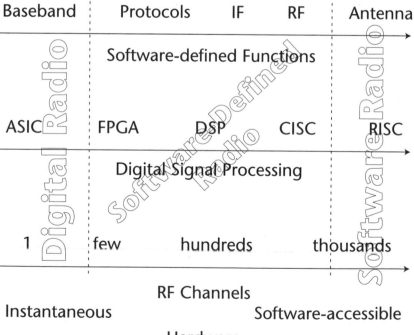

such an ideal implementation may not be practical because of size and power consumption. As processor technology advances, however, air interfaces that require Application Specific Integrated Circuits (ASICs) today may be implemented on general-purpose processors in the future.

The software-defined radio (SDR) therefore compromises the software radio ideal, shown in Fig. 1 in order to implement practical high-performance devices and infrastructure with current technology. SDRs are implemented using an appropriate mix of ASICs, Field Programmable Gate Arrays (FPGAs), Digital Signal Processors (DSPs) and general-purpose microprocessors, with reduced instruction sets (RISC) or complex instruction sets (CISC). This migration from digital radio towards the ideal software radio, also shown in the figure, is a work in progress since 1991 that may take another decade or more. The architecture and theory of the SDR migration from digital to software radios are becoming well known.[2]

In addition, the convergence of communications and computing, along with Moore's Law, enables increasingly compact multimedia SDR, with many embedded non-radio functions as illustrated in Fig. 2. In this model of radio functions, sources such as users and networks offer voice, data, and multimedia streams to source coding and networking services of the SDR. Information security (INFOSEC)—such as GSM authentication and stream ciphers—protects these services from unauthorized access. The modulator/demodulator (modem) converts source-coded bitstreams to the channel code, which is upconverted to intermediate

Figure 2. Converged SDR-PDA Functions

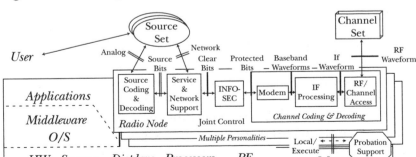

(IF) and carrier (RF) bands. These radio services and multimedia applications share open-architecture middleware and computing services of a heterogeneous processor platform. Multiple "personalities" enable waveforms and multimedia services such as phone book, calendar, photo galleries, maps, and the like. Radios may use such computational models to mediate downloads of new personalities and to test them before enabling them.

Looking toward the creation of SDR products, the global SDR Forum and the U.S. Department of Defense (DoD) Joint Tactical Radio System (JTRS) Joint Program Office (JPO) together defined an architectural framework, the Software Communications Architecture (SCA).[3] This object-oriented model incorporates the functions of fig. 2. Its applications factory supports evolution, constructing waveforms from components. XML defines interfaces among classes of radio objects. The SCA is now being used to develop the first DoD SDRs, the U.S. Army's JTRS Cluster 1 vehicular radios.[4] The JPO lists the types of air interfaces as "waveforms" on their website, as follows: "The Prime System Contractor will provide software waveforms required for Cluster 1, and ensure all waveforms are ported to both JTRS radios and to the Waveform Test Environment designated by the JTRS JPO. Waveforms being developed by JPO under Cluster 1 include EPLRS [a tactical data network], UHF SATCOM, HF (ALE & ECCM) Air Traffic Control (UHF Voice, Data; HF Data) VHF/UHF LOS [Line of Sight] (VHF AM, FM; UHF FM/PSK/CPM), Have Quick II [an Air Force waveform], SATURN, UHF LOS (Link 4A, Link 11B), UHF DAMA SATCOM, LINK 16 [JTIDS], LINK-11A, [and] WNW. In addition, the JTRS JPO is developing the SINCGARS ESIP [an Army waveform], APCO-25, IFF Mode S, and Cobra waveforms under separate acquisition efforts. These waveforms will also be available for Cluster 1 sets".[5] This is a wide range of conventional voice and data modulation types across spectrum bands from 2 MHz to about 2 GHz.

Commercial SDRs include handsets and software-defined commercial base stations. Handsets may be called wireless Personal Digital Assistants (PDAs) or multifunction cell phones, depending on the radio technology and market segment. Computer-communications convergence is the major trend blurring the technical distinctions. Prior generation baseband-DSP (digital signal processing) base stations included more than one hundred analog IF links from the analog RF converter to one hundred baseband ADCs and DSPs. In GSM digital base stations, the separate Transcoder and Rate Adaptation Unit (TRAU) convert GSM's 13 kbps voice to 64 kbps DS0 PCM. SDR base stations now convert IF to wideband digital, for example, with a 70 MHz ADC, accessing the entire 25 MHz GSM allocation with a single device. Digital IF filters replace analog IF filters, and digital signal distribution replaces analog signal distribution. Greater processing capacity of the consolidated baseband DSPs subsume the previously stand-alone TRAU. IF digitization, digital filtering, and digital signal distribution improves the reliability of such base stations, which began shipping in 2000 and 2001. An SDR base station may now be configured in a sports utility vehicle (SUV), for instance, to reconstitute and/or supplement communications in a humanitarian operation.

B. Aware-Adaptive (AA) Radio

Awareness radios, defined by Blust in September 2003, are those radios that have a capability to be aware of something, typically sensing it in themselves or in the local physical environment. Awareness includes extracting information useful to a user such as location or temperature. It may also include "knowing what it knows," such as "knowing" that its video port can take a digital snapshot or video clip. The difference between a radio that "knows" and a radio that is merely capable is that the aware radio employs a computational ontology, such as the Radio Knowledge Representation Language (RKRL), an organized set of terms over which it is capable of conducting a meaningful dialogue, if in a computer-oriented language such as an XML dialect. The awareness radio has the internal capacity to sense the thing of which it is aware locally, unaided by an external entity (that is, with no help from the network, the user, etc.) It knows what it knows because it autonomously refers its actions to its computational ontology. For example, a radio equipped with a GPS (global positioning system) could sense its coordinates, displaying them to the user, who then is aware of the exact location. The awareness radio might look up the coordinates in an internal database, refer them to its computational ontology, and infer that the radio itself is now in the location of a current disaster area. GPS alone doesn't make a radio location-aware. The combination of location sensing and reference to a computational ontology enables the radio to "know what it knows."

The Semantic Web currently provides major impetus for open computational ontologies combined with semantics-enabling referential logic systems.[6] Such ontologically based reasoning systems enable any computing entity to "know what it knows" in that it can use the ontology to set context for a discourse. The word *disaster*, for example, carries a wealth of different semantics. In a computer context, it could have more to do with backing up data than with helping people with medical needs. Standard ontologies permit both people and computers to attribute common meanings to terms, enabling a computer to know what it knows.

This standardization of terms is particularly important in the design of radios that can be effective with minimal human intervention in disaster relief settings. Adaptive radios, as defined by John Chapin of Vanu Inc. in September 2003, are those radios that can adapt their behavior, either as tasked by a network or as required given the local physical and RF environment, invoking an appropriate preprogrammed or network-defined behavior. Adaptive radios can "do what they are told" by a network, based on a preprogrammed behavior. The behavior of an SDR could be preprogrammed at the factory, or it could be a new one downloaded over the air (OTA). In either case, the behavior has been sanctioned for use on that specific class of device, such as approved by a regulatory authority and certified by an OEM (see Harada's bitmap concept for keeping track of such regulatory approaches.)

If a radio operates on only one home network, then the semantics of its operation are defined at design time, and there is no need for computational ontologies. But the disaster-relief SDR is meant to operate on a multiplicity of networks, so semantics must be agreed to across networks and nodes in real time as the scene develops. Such contingencies motivate the Aware-Adaptive (AA) SDR (synthesized from Blust and Chapin by Mitola in September 2003). These radios adapt their behavior based on their awareness capabilities. A radio can be aware of location but not adapt to it, just providing location information to the user. It could adapt its spectrum use based on location, such as transmitting on U.S.-approved spectrum while in the United States and switching to the European-approved spectrum for the same function in Europe. The scope of topics of which an AA radio may be aware is suggested in Fig. 3. In this particular form, the ontological knowledge is displayed as a loose hierarchy of concepts related to classes and states of users. RKRL, on the other hand, is more spatially oriented, capable of defining all the things in the universe that extend over space and time as well as related abstractions. The IEEE Standard Upper Ontology (SUO) and the semantic web ontologies may offer more robust treatments of general knowledge suited for AA radio development.[7] Use cases illustrate how AA radios might work:

1) AA Radio Disaster Relief VHF Radio Policy Use Cases. It is easy for an SDR to be adaptive to a topic without being aware of that topic. For example, it could be told to change its behavior (to "adapt"

Figure 3.   Aware-Adaptive (AA) and Cognitive Radio (CR) Build on Awareness

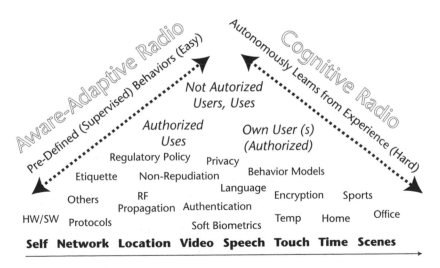

in the large sense) by a network without having related local sensing. A cell phone that is told to not use certain channels of its
VHF mode by the network might not be able to figure out on its
own that police are using that channel in the specific jurisdiction
and at that particular time.

Other more computationally aware SDRs might interpret the
NeXt Generation (XG) policy language being developed by
DARPA, broadcast on an "advice" channel.[8] The XG policy-aware
radio could avoid the channels in police use in VHF by interpret-
ing and adhering to the broadcast policy specifying channels in
use by the police. XG AA radios would not have been explicitly
preprogrammed to notch out these VHF radios by a host network.
In fact, there need be no host network, just the local XG policy
broadcast.

Still more sophisticated AA radios could recognize the push-to-
talk behavior pattern of the police radios and politely back off of
those channels. In other words, the XG policy channel could
specify that push-to-talk police radios are in use in the area on
certain large bands, employing specific call signs, or data formats,
without specifying which channels to avoid. The AA radio would
then monitor channels before use, recognizing, say, the push-to-
talk behavior and possibly detecting the call signs, for example,
using advanced versions of the speech recognition technology
now in use for 800 directory assistance in the United States. Such
general classes of radio "etiquette" have been described in the
literature.[9] These behaviors seem particularly helpful for humanitarian operations in which the players must infer the rules for
radio spectrum usage as the situation evolves over time.

2) AA User Awareness Use Case. Suppose the AA radios in a disaster
are user-AA. Each user-AA radio recognizes its own user, such as
by voiceprint or recognizing the 3D video face or both. Today, in
order to protect your personal information from others, you need
a secure operating system and a PIN. I happen to be typing this
chapter while traveling on an airplane, and it's pretty easy for the
person sitting next to me to observe my PIN if I'm not careful.
But if my AA radio (or laptop) could actually recognize me, my
face and voice, then I wouldn't worry as much about somebody
stealing my PIN on the way to the disaster area. They'd also have

to "steal" my face and voice in order to use my PIN on my user-aware AA wireless laptop. That might be a good thing if somebody were to steal it, as someone did six months ago. Did they get my personal data? Did they steal my identity? I may never know. In the disaster area, I might introduce other users to my radio, instructing it to assist those users as they would assist me. If my radio is somehow lost, it could describe me to others, perhaps assisting them if they meet criteria I have specified in advance, ultimately finding its way back to me, but contributing to the relief effort in a more flexible way than a conventional radio.

3) SDR Evolution to AA-Enabled Cooperation. Not all SDRs are aware of users or spectrum-use policy. For example, a point-to-point two-way SDR used by police might not be aware of anything. Such a police SDR could become network-AA by uploading a new radio personality, say via CD-ROM from an FCC certified supplier. Its new VHF radio AA personality could be network-adaptive by courteously responding to network advice offered by other AA radios on an ad-hoc VHF "advice" channel. In a disaster relief deployment, the radio could learn, for example, that "VHF channels [nearby] are using Part 15 low power mode for an ad-hoc RF LAN" (that is, within the zone in which the advice is broadcast). The advice might be broadcast in XG policy language or in some other dialect of XML. The radios could share definitions via SUO or a semantic-web ontology, for example, of disaster-related terminology. The newly educated polite AA radio would not use that channel in that zone. Instead, it would use other VHF frequencies to establish connectivity to distant nodes from the vicinity of the declared RF LAN zone. Thus, relief agencies arriving in a disaster area could upload AA personalities to their SDRs so that the radios could cooperate effectively at the site, semiautonomously declaring spectrum use detail and relieving the relief workers to perform search and rescue, give medical attention, and the like. Thus SDR evolution to situation-aware AA radios would take less time and fewer people for the establishment and control of the enabling RF networks.

4) Social Networks Use Case. Let's say the disaster-relief SDR is user-AA. When a group forms to distribute food, their AA radios introduce themselves. The radios generate an ad-hoc identifier for their group, say Water-Boys-1. The realization of this vision may be years in the future, so by then, the radios would be able to scan all the VHF channels in the band and automatically keep track of the other members of their ad-hoc SDR group. If other relief workers approach, their group name, which might also be Water-Boys-1, wouldn't match a random group ID code assigned by the radios when introduced, so the radios could advise their users "There's another group of relief workers nearby calling themselves Water Boys—would you like to let them know of your shared interest?" If not, then the radios could either scramble their name (that could be the default if more privacy is desired up front) or could welcome the social interaction and maybe the assistance enabled by the user-AA social networking feature.

5) Privacy Enhancement Use Case. Such aware-adaptive radios, like all other transmitting devices regulated by the FCC, would be certified by their manufacturers to obey all applicable regulatory mandates. One requirement for AA radios might be that all such radios be policy-aware and policy-adaptive. This means that an AA radio would read spectrum use policy from the host network or from some local broadcast (for example, provided by data under FM broadcast the way broadcast stations identify themselves to smart radios in automobiles today). The manufacturers would have some responsibility to see that radios obey policy and that older radios that cannot obey new policy at least do no harm in ignoring it. For example, the older radios might not be as agile in the use of spectrum, but the newer radios would "be polite" to these older radios, working around them in the same way that cars avoid pedestrians. Thus, it is easy to see how technology could over time come to support the development of AA radios consistent with both current FCC rulings and future changes.

A policy-AA radio might be told to "not be agile" anymore if a liberal regulatory policy were subsequently reversed. Policy-AA radios could obey any policy whatsoever. For example, one could craft privacy policies. Until I gave it to the Salvation Army last year, I had an old Bearcat scanner. I bought it in the '80s to listen to pit crews at NASCAR races. When the FCC changed the spectrum allocation, my radio was unaware of that, so it let me listen in on cell phone conversations. With hardware-based radios, that's just too bad, and the onus is on me not to listen, which I didn't, of course. But with a policy-AA radio, the policy broadcast channel would have advised my policy-AA SDR of the list of channels my scanner is allowed to scan in VHF/UHF. Those cell phone channels would not be on that list, so a policy-AA radio would not let me listen to those channels. Thus, the policy-AA technology has the potential to enable better compliance with privacy rights than either SDRs or hardware-defined radios. Nothing says a user must actually go to the supplier or download a new personality for a non-policy-AA SDR that will take scanner channels away. These classes of technology require legal adjuncts to hold the user responsible for behavior inconsistent with policy. A policy-AA radio, however, is constantly getting and obeying the new rules, whatever they may be.

The ways in which radios and wireless PDAs may become adaptive and aware is limited only by the imagination. By specifying exactly what the radio is (locally) aware of and exactly how it can adapt (either to local awareness or as commanded by a host network), suppliers can more clearly communicate with potential users about what "smart" radios can and can't do.

C. Cognitive Radio (CR)

Cognitive radio, in short, is an AA radio that can learn from experience. In the literature, the term first signified a radio that employed model-based reasoning to achieve a specified level of competence in radio-related domains.[10] Subsequently, CR architecture was defined to include natural language processing and integrated machine learning.[11] Recently, CR have been described that employ game theory for spectrum sharing.[12] The ontology-based radio step towards CR has also been described.[13]

The CR architecture investigated at KTH for CR1 employed the cognition cycle illustrated in fig. 4. Control cycles through the stages Observe, Orient, Plan, Decide, and Act, with integrated machine learning. The outside world provides stimuli. Cognitive radio parses these stimuli in a perception hierarchy to recognize the context of its communications tasks. Incoming and outgoing multimedia content is parsed for the contextual cues necessary to infer the communications context (for instance, urgency). Thus, for example, the radio may infer that it is going on a trip to a disaster area (with some probability) if the user E-mail to a travel agent specifies a destination located in a foreign country where (the radio knows from ontology processing of news) a disaster event recently took place. The Orient-stage determines the urgency of the communications in part from these cues in order to reduce the burden on the user. Normally, the Plan-stage generates and evaluates alternatives, including expressing plans to peers and/or the network to obtain advice. The Decide-stage allocates computational and radio resources to subordinate (for instance, SDR) software. The Act-stage initiates tasks with specified resources.

In CR1, stimuli are organized into increasingly more comprehensive perceptual structures. Language stimuli in CR1 are structured into words, phrases, dialogues, and scenes. Visual stimuli were not implemented in CR1, but the architecture provided for the association of objects in scenes with spoken dialogue and other auditory and sensory stimuli (including temperature and acceleration, among others) for the capture of more or less complete multimedia scenes. CR1 remembered all stimuli, discarding nothing and accessing its "lifelong" stimuli efficiently using Java hash-maps. Although CR1 didn't "live" very long, all the stimuli of a typical one hundred-year human life could be stored in approximately 100 terabytes, which seems large until you think that just a few years ago the idea of storing 100 gigabytes on a laptop seemed far-fetched. If Moore's Law continues unabated, within twenty years we will have the 100-TB laptop, so the era of "store everything" is not far off. Each stimulus therefore is either known or new. Known stimuli are conceptual anchors for a case-based reasoner to match a current partially known scene to the nearest prior experience. CR1 exposed research issues of highly integrated machine learning.

Cognitive radio is a goal-driven framework in which the radio autonomously observes the radio environment, infers context, assesses alternatives, generates plans, supervises multimedia services, and learns from its mistakes. This observe-think-act cycle is radically different from today's handsets that either transmit on the frequency set by the user or blindly take instructions from the network. Cognitive radio technology thus empowers radios to

Figure 4.   The Cognition Cycle

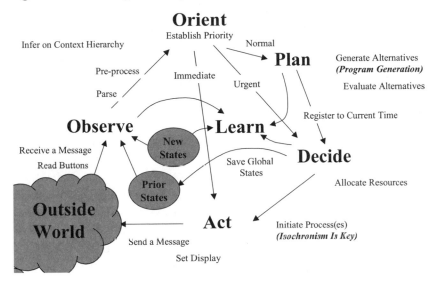

observe more flexible radio etiquettes than was possible in the past.

One of the key research issues to surface in CR1 is the degree to which training influences the behavior of a cognitive radio. Since CRs learn from experience, autonomously matching the current scene with prior knowledge and autonomously deciding what to do based on feedback from the environment, primarily the user, the behavior of a CR is not very predictable. Somewhat like forecasting the weather, the envelope of behavior of a CR with a given level of training may be limited. But over time, as machine learning research has demonstrated, performance may become brittle, and unexpected behaviors or inability to learn further may result. Reasoning with explicit representations of un- certainty and various unsupervised and supervised methods of constraining behavior assist in circumscribing the behavior of possible CRs, but in general, a CR's learning ability makes it a "Turing-capable" system, which, through Gödel's famous proof, is provably inconsistent. The potential for inconsistency raises the specter of learning unsanctioned behaviors, so until CR is better understood, it seems wise to pursue research but perhaps not un- leash them on the public. Thus, in the sequel, the potential bene-

fits of SDR to humanitarian operations are characterized through
AA radios, leaving the CR that can learn and adapt from its own
experiences in the category of DARPA-hard research issues, not
soon-to-be-released commercial (or military) products.

In the sequel, this paper develops the design of an SDR for
humanitarian operations, illustrating key benefits that accrue by
enabling the use cases.

2. AA SDR for Humanitarian Operations

Consider a mobile communications capability for disaster relief.
A comprehensive capability would include mobile SDR infrastructure, AA vehicular nodes, and AA handsets. System designs may
employ open architecture such as the SCA. Architecture consists
of functions, components, and design rules. The SCA defines
components at such a high level of abstraction that one needs
illustrative implementations to characterize the potential benefits
of the architecture.[14]

A. *Humanitarian Operations Disaster Relief Scenario*

Ibrahim Osman defines a disaster as a situation or event that over-
whelms local capacity, necessitating national or international assistance.[15] He groups natural disasters (avalanches, famines,
floods, forest fires, and the like) with technical disasters (for example, industrial or transport accidents). He differentiates these
naturally occurring events from those "complex disasters" in
which there is a complete breakdown of authority resulting from
internal or external conflict and which meets other criteria. According to Osman, between 1994 and 2001, about U.S.$154 bil-
lion in relief was afforded to natural disasters globally.

To envision SDR benefits, suppose a medium-sized urban area
(fewer than 100,000 people) has been decimated by a natural
disaster. As illustrated in Fig. 5, the disaster area has been largely
obliterated. Assume that some fraction of the populace has en-
joyed the use of cellular telephone but that the disaster has seriously impaired the wireless network. At the periphery of the
disaster area, connections are available to the fiber and/or micro-

Figure 5.   Humanitarian Relief Example

wave access to the global public switched telecommunications network (PSTN).

Further assume that an appropriate national or international authority would like to acquire a capability to rapidly reconstitute communications in such disasters. Sample customers include the U.S. Federal Emergency Management Agency (FEMA), the European Community (EC), the United Nations (UN), a G7 government, or a non-governmental [international] organization (NGO) like the Red Cross/ Crescent. In many situations, numerous local, state, and national institutions converge on the disaster area along with NGOs and international relief teams to search for survivors, supply medical assistance, set up temporary shelters, prevent crime, and reconstitute the necessities of life. For military relevance, international assistance may include the deployment of military units, for instance, to assist local police in maintaining civil order.

To frame the discussion of SDR, consider the conventional approaches to supplying necessary communications. In developed nations, dozens of local, state, and federal organizations arrive at the scene with voice radios operating on different bands and modes, based on low-cost purchase of handsets and vehicular radios optimized for local, regional, or specialized nations use. A communications director aggregates one of each radio type at an ad-hoc Radio Communications Facility (RCF) improvised on high ground. People staff the RCF, bridging calls manually across the plethora of bands and modes. This conventional approach requires no capital outlay and can be done anywhere and at any time, but the victims may not be well served because of the attendant time delays and compounded communications problems overcoming radio interoperability, symbols, jargon, and language barriers. This leads one to more seriously consider the potential benefits of an open-architecture SDR approach to communications services.

B. SDR System Needs

Initially, communications systems engineers may think of SDR as an efficient means of setting up a conventionally staffed RCF. A conventional needs analysis establishes the relationships among radio system functions, components, design rules, and costs. Systems level communications needs for a disaster relief system are summarized in Table 1.

The answers to the needs questions define the top-level requirements for an SDR. As with other wireless services, physical area per cell and total numbers of subscribers are first-order determinants of the traffic offered to the system. Each subscriber offers some mix of traffic in order to obtain the indicated information services wirelessly. The fundamental measure of voice traffic is the Erlang, the international unit of traffic intensity that represents an average of one circuit busy out of a group of circuits.[16] Wireless infrastructure provides capacity in Erlangs per square km at a given Grade of Service (GoS)—the probability that a connection is made—and for a given Quality of Service (QoS)—the data rate, time delay, and packet loss parameters of the circuit or data network. In the postulated disaster relief case, there are four major classes of subscriber: police, fire-rescue, local populace, and National Guard. Each class brings its own indigenous vehicular and handheld radios and wireless PDAs. These radios establish the radio bands and modes that must be supported by the SDR mobile disaster relief infrastructure. In addition, those people who are providing the communications services

Table 1.   Illustrative Needs Analysis

Needs Questions	Illustrative Answers
Physical Area?	3–5 local areas of 2–10 km radius each
Classes of Subscriber?	Police, Fire, Rescue, Local Populace, National Guard
Numbers of Subscribers?	10–20 Local and/or National Police Agencies 20–100 Fire and Rescue Squads (10 Helicopters) 50,000 Local Populace (including 20 Light Aircraft pilots) 500–3,000 National Guard Troops with 20–50 Aircraft
Information Services?	Core: Voice, E-mail, Tasking/Scheduling, Databases, Fax Growth: Video teleconferencing, Telemedicine
External Interfaces?	Network: T/E-1 to T/E-3 SDH (Microwave, Fiber), SS7
Cost?	"A few million dollars"

also need local communications. Call these the organization and control (OC) users.

A formal needs analysis examines the general scenario by generating a variety of use cases, for example, by using a modeling tool like UML (Unified Modeling Language). The needs of OC users may be derived by creating formal use cases, detailed vignettes that force one to think about significant details of the application. OC users might need a geospatial information system (GIS) to visualize the distribution of the entities in the network. A discrete event simulation can be used to characterize queuing delays of message traffic needed to support the E-mail, scheduling, and database services (for instance, OpNet). In addition, UML simplifies some aspects of use-case analysis. UML's use-case view keeps track of external and internal actors and forces one to push through the tedium of each use case.

The needs analysis for an SDR-based product may attempt to

limit the needs so that the complexity of the SDR software is mini-mized. Typically over half of the cost of developing an initial SDR product is in the software. To limit the services provided is to limit the software complexity. Initial SDRs for humanitarian operations may have limited awareness and limited adaptability. Some de-gree of location awareness, user awareness, and spectrum aware-ness seem beneficial to virtually any humanitarian operation if the SDR products could implement the use cases. To see how this would work, examine the radio resources in greater detail.

C. Disaster Relief Radio Resources

A top-down analysis of the disaster-relief case study identifies the communications resources. Each class of participant entails radio equipment and rights to use the radio spectrum. The potential radio resources for disaster relief are illustrated in Table 2.

This first level analysis yields a range of numbers of radio units that will be brought into the disaster area. Each vehicle that car-ries radio equipment is referred to as a radio node. Each node has the potential to access its native allocated or licensed spec-trum. Some nodes will have the capability to cover multiple bands outside of their normal bands of operation. In order to provide a mesh of connectivity in the disaster area, there must be both some degree of overlap of radio access and some baseband switching capability.

D. Requirements Analysis and Top Level Trade-offs

Illustrative requirements for mobile disaster-relief infrastructure Humanitarian Operations Base Station (HOBS) are provided in Fig. 6. The SDR platform is configured to the RF bands and modes, maximum number of subscribers, and services to be pro-vided. In addition, the mobile radio equipment must fit in a mo-bile vehicular platform. As a design goal, each HOBS node would be configured in a commercial four-wheeled SUV. This SUV may be equipped with an electric or hydraulic mast perhaps ten-meter maximum height.

1) Provisioning. The number of SUVs may be decided analytically

Table 2.   Humanitarian Operation Radio-Related Resources

Parameter	Aspect	Potential Resource
Physical Area	3–5 areas of 2–10 km radius	3–5 radio cells; 18–150 sq km total area
Classes of Subscriber	Police, Fire-Rescue, Local Populace, National Guard	APCO radios; cell phones; military radios, wireless trunks, and switches
Numbers of Subscribers (by Class)	10–20 Police Agencies	10–20 command nodes (APCO/ Tetra) A few special radio types (e.g. U.S. FBI)
	20–100 Fire and Rescue with 10 Helicopters	20–100 vehicular nodes + 100–1000 hand-held 10 air mobile radio nodes (3 or more radios each)
	50,000 Local Populace including 20 Light Aircraft pilots	500–10,000 cell phones, 500–3,000 cordless 20 light air mobile nodes (2 or more radios each)
	500–3,000 National Guard Troops with 20–50 Aircraft	50–300 squad radios, 12–80 company radios, 3–10 high level command network radios, radio relays 20–50 air mobile radio nodes (3 military radios)
Classes of Information Services	Voice E-mail —Tasking/ Scheduling, —Databases Fax Video teleconferencing Telemedicine	Isochronous narrowband traffic Unformatted messages (rescue, local, victims) formatted (requires client software) formatted (requires client and server) Hardware or software sources Isochronous MPEG traffic Isochronous wideband traffic
External Interfaces	Network: T/E-1 to T/E-3 SDH (Microwave, Fiber), SS7	Fiber or microwave interface to the PSTN

based on GoS, Erlangs of traffic offered per subscriber, and spatial area covered by the ten-meter mast antenna. Alternatively, one may derive the number of nodes top-down through similarities and differences with the reconstituted cellular service. Typical cell sites support about one hundred subscribers. If a mix of VHF/UHF, HF, and cellular subscribers is envisioned, then there might be one hundred to two hundred potential subscribers in each of the two major bands for a total of two hundred to four hundred users. The peak capacity of each van could then be set between one hundred and four hundred. The lower the capacity, the lower the cost of the system. Consider each of the additional requirements in turn.

RF bands are selected for interoperability among emergency teams. LVHF is required for interoperation with National Guard using SINGCARS. VHF and UHF are needed for communications with aircraft (including the military Have Quick mode) and for federal, state, and local law enforcement push-to-talk radios. HF was not explicitly called out, but if an emergency occurs in a mountainous region, HF Near Vertically Incident (NVI) skywave is an effective way of connecting teams operating in adjacent mountain valleys. HF is therefore included.

In addition, the restoration of the cellular telephone service requires UHF modes in the 850–900 MHz band, as well as PCS

Figure 6.   Summary of Requirements of HOBS

- RF Bands
  ☐ HF   ☐ LVHF   ☐ VHF
  ☐ UHF (.3 to 3 GHz)   ☐ SHF
- Instantaneous Bandwidth
  ☐ CDMA __ MHz
  ☐ Service Band __ MHz
  ☐ Agility Bandwidth __ MHz
- Sensitivity & Near/Far
  ☐ __ dB Near Far
  ☐ 0 dBm Cochannel
    Inferference
- Multi Standard Support
  ☐ Analog   ☐ GSM/TDMA
  ☐ JTIDS   ☐ EPLRS-like

- Maximum Subscribers: N = __
  ☐ Remote BTS   ☐ Cell Radius
- Services
  ☐ Voice   ☐ Data (IP Level)
  ☐ Video Teleconference
  ☐ .008—2 Mbps
  ☐ Voice Mail   ☐ Talk Dial™
  ☐ CORBA
- Network Interfaces
  ☐ SS7
  ☐ __ SDH or __ ATM
- Platform
  ☐ 1 Small Van

modes to 2500 MHz. Such networks can be very complex. CDMA networks, for example, entail more than one thousand control parameters, so the mere implementation of a CDMA waveform does not fully realize a viable CDMA network. In isolated tests, mobile CDMA base stations have been deployed with default parameters to provide local area wireless connectivity without the benefit of extensive network infrastructure. This stand-alone mode is postulated for HOBS.

2) Cross-Connect. An additional reason for RF coverage is the need to cross-link HOBS SUVs to each other efficiently. Suppose two vans are ten miles apart, each supporting one hundred local users. Some fraction of these users will need to communicate with users supported by the other van. The choices include HF, fiber, VHF/UHF, satellite communications, and SHF point-to-point radio relay. HF may provide the connectivity for a small number of channels in rugged terrain. But if on the average there will be twenty to thirty calls between the two vans, T-1 (twenty-four channels) or E-1 (thirty channels) Line of Sight (LOS) microwave service is warranted. With ten vans, the cross-connect traffic may be two hundred to three hundred channels. Although it is possible to lay field fiber, this mode is subject to breakage, especially in disaster situations. Terrestrial microwave easily provides T-1 to T-3 capacities with relatively modest bandwidths and subsystem complexity. For this example, SHF in the 4, 6, or 11 GHz microwave bands would be the high capacity cross-connect mode.

3) PSTN Interoperability. The subscribers must also be connected to the PSTN. In some areas, the PSTN may employ SHF microwave to protect primary fiber infrastructure. But most service providers in the United States now protect ("back up") fiber with other fiber paths. So the vans should have a fiber interconnect port compatible with the SDH and SS7 for interoperability with the PSTN. Since the design of such interconnect is not central to software radios, the sequel will reflect the assumption that the physical interconnect and the necessary driver software is available as commercial off the shelf (COTS) products. The software radio will have to deliver isochronous streams to the interface and route streams from this interface to radio users. But the design of the interface itself is not central to the software radio.

4) Digital IF Trade-Offs. Instantaneous bandwidth, sensitivity, and dynamic range (near-far ratio) of each radio band are driven by the commercial standards. Most state and local police and fire and rescue units employ push-to-talk VHF/UHF AM/FM radios. The instantaneous bandwidths range over the set {4, 8 1/3, 12.5, 25, 50, 100} kHz. Commercial cellular standards, on the other hand, include the IS-95 CDMA system with its 1.2288 mchip/second spreading rate with 1.25 MHz bandwidth and W-CDMA with 5, 10, and 20 MHz instantaneous bandwidths. GSM requires only 200 kHz of instantaneous bandwidth per burst, but the FH modes can hop over 25 MHz. SDR implementation of the FH mode requires 25 MHz bandwidth for transmission and reception. This drives the DAC and ADC requirements. In addition, the dynamic range is set by the near-far ratio, for instance, 90 dB for GSM. The details of these trade-offs are addressed in contemporary software radio texts.[17]

5) Digital Signal Processing Trade-Offs. The tradeoffs of digital IF and choice of digital signal processing (DSP) interconnect and processors are driven in part by the design of the digital IF. Specifically, the first stage of digital filtering must be consistent with the sampling rate and dynamic range of the digital IF. Wider digital IF bandwidths drive one towards applications-specific integrated circuits (ASICs) and/or field programmable gate arrays (FPGAs). Narrower bandwidths and thus lower data rates and less demanding digital filtering may be more consistent with DSP chips or multimedia-capable general purpose processors. The increased programmability of the more general purpose devices is illustrated in the trade space of Fig. 7, Software-Defined Radio Trade Space. In the figure, point A refers to the HF STR-2000, a product of Standard Marine AB developed in the mid-1990s using dual Texas Instruments C30 DSPs for bandwidths on the order of 10 to 20 kHz, which for HF voice and Morse code was a reasonable IF. For the higher RF bands (VHF, UHF, and SHF), this bandwidth would be considered baseband, which is where it is positioned in the figure. Point B of the figure illustrates the typical commercial off the shelf (COTS) cellular radio handset of the late 1990s with ASIC CDMA de-spreader and baseband DSP to yield a point between the ASIC and FPGA for weighted average

Figure 7.   Software-Defined Radio Trade Space

Digital Access Bandwidth: "ADC at the Antenna" -Requires Wideband Analog Front Ends

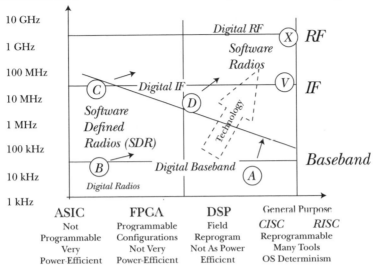

Programmability After Deployment / and Power Efficiency/

DSP. Point C shows that the SDR cell sites deployed in 2000–2003 digitized typically a 25 MHz spectrum allocation, selected subscriber channels with ASICS, and processed the subscriber basebands using DSPs. Point D illustrates DARPA's SPEAKeasy II software radio prototype. This design aggressively emphasized software reprogrammability and therefore had few ASICs, instead delivering most processing capacity via commercial DSPs and other programmable chips (such as multiply-accumulator co-processors).

Point V occurred in November 2003 with the field trials of the first commercial cellular base station to use COTS Intel Pentium processors and the LINUX operating system to deliver GSM service in Texas by Vanu, Inc.[18] Point X represents the Ideal Software Radio (SWR), a point on the horizon that is useful for characterizing the degree of flexibility and programmability of software-defined radios. That point may never be realized cost-effectively. But plotting SDR evolution towards that point illuminates the pace and nature of SDR product evolution.

6) Enabling Services for Humanitarian Action. As processing capacity costs decrease with Moore's Law, the affordability of SDR base stations configured in SUVs for HOBS continues to improve. These improvements enable services beyond voice and data. Compressed digital image exchange entered the commercial mainstream in 2003. HOBS class mobile nodes can be configured with compressed digital video using data rates of 64 to 128 kbps. With some quality of service (QoS) guarantees for the control of remote instruments, this enables two-way telemedicine. RF tags applied to patients combined with simple time-difference of arrival (TDOA) emitter location algorithms permit HOBS to track the location of patients, doctors, and emergency equipment as all move within the humanitarian operations area. RF LANs set up ad hoc by HOBS using ISM bands can enable much needed record keeping with data capture at point of origin via PDAs. Even pagers from home nations can be accommodated by downloaded waveforms if the host government authorizes such innovative use of the radio spectrum. To keep up with the dynamics of people with different roles using the same PDAs, handsets, and base stations as others, SDRs for humanitarian action may employ "soft" biometrics to identify users. Soft biometrics include images of faces and voice recognition. With appropriate controls to respect privacy policies, such tools enable rapid adjustments in priorities for the use of the radio's physical layer, which is always in contention in busy environments. The chiefs of police, for example, may move from one base station to another and at any given time may need to switch from their own handset to some other handset as the situation dictates. Soft biometrics permit the radio to recognize that when the new user says "This is the Eagle One," this is the chief of police, and he needs to preempt other users for access to the physical layer. Other language tools, including real-time language translation services and near real-time multilingual graphics annotation, are also enabled by HOBS spare processing capacity. Specifically, a system like HOBS needs to be provisioned with about 50% spare processing capacity in order to guarantee real-time performance of isochronous voice streams. This creates about 20% spare capacity that can be used for language processing when resources are available because voice, data, and video traffic loading is low. That capacity smoothly shifts to isochronous threads when traffic volumes increase. Some users would detect some delay in translation or annotation services, which can also be a function of policy-defined user priorities for communications services.

E. Illustrative HOBS System

To complete the design, each of the areas listed in fig. 6 must be analyzed in detail. Texts develop such architecture trade-offs further. This brief treatment summarizes a representative design as follows.

1) *Communications Services.* The HOBS mobile infrastructure system consisting of five or six SUVs configured with SDR would integrate the communications capabilities of diverse police, fire, rescue, NGO, and military organizations. In addition, it will bridge communications of national and international relief agencies with disparate communications equipment into the local disaster-recovery operations. Finally, it will integrate military and National Guard communications.

Core services would consist of voice, data, and video telemedicine. Voice services could include voice mail. A language-based Talk-Dial ™ (LTD) capability could access rescue workers by functional category based on real-time speech processing and directory management. A rescue worker could ask the SDR for "the nearest person with a breathing apparatus" if the call sign or telephone number were not known. Like a helpful operator, LTD would connect the two team members. Rescue personnel therefore need know only the name of the person or the general category of function in order to get the right person. A HOBS system manager would assign a virtual telephone number to each relief team member and then track that participant's location and communications mode for improved connectivity. Data services include wireless E-mail. In addition, HOBS establishes a gateway to the PSTN via microwave and fiber optics.

HOBS as an SDR thus bridges communications across both radio modes and language barriers, reducing confusion and enhancing team efficiency. Since each HOBS SUV is equipped with a ten-meter mast, it establishes a cell within which commercial cellular handsets (for instance, of the victims) can operate even when local cellular service has been interrupted by the disaster conditions. Organic radios of disaster support teams connect to the local HOBS van for bridging to proximate or distant team members. They obtain frequency and mode assignments for local communications from the spectrum management policy software in HOBS.

2) RF Bands and Modes. HOBS services HF, LVHF, VHF, UHF, and SHF. HF AM and ALE provide voice and data circuits using NVI modes in mountainous regions. LVHF coverage integrates contributions of military and National Guard units. VHF coverage of the 100 MHz air traffic control band permits coordination with aircraft and the reconstitution of communications at an airfield. VHF/UHF push-to-talk AM, FM, and TETRA digital radio modes are supported in all bands, subject to frequency coordination with the HOBS spectrum managers. Additionally, UHF cellular coverage includes 1G, 2G, and 3G air interface modes. The RF LANs operate on the 2.5 MHz ISM band so that wireless laptops can be used in the vicinity of the HOBS vans for status monitoring and coordination displays. Telemetry modes permit HOBS to uplink patient status data via wireless and PSTN links to remote medical personnel. Streaming video supports telemedicine. Switching of voice channels is accomplished in software under the control of ETD. The interface to the PSTN employs SS-7 and SDH Levels 1, 2, or 3 trunking through microwave or fiber optic media.

3) Capacities. The HOBS design supports two thousand emergency personnel per node with up to five vans. The internal capacity of each van is two hundred Erlangs of traffic. Band coverage consists of ten subbands from six antenna channels.

F. Illustrative Components

HOBS includes hardware and software components as follows:

1) Hardware Components. An illustrative hardware design is illustrated in fig. 8. HF supports a 6 MHz subset of the HF band, tunable between the LUF and MUF. LVHF is accessed in parallel using a 150 MHz ADC. This limits the near-far ratio to roughly the dynamic range of the ADC, less typically 6 dB for automatic gain control. There are three tunable subbands in the VHF range from 88 to 400 MHz. The low subband would access commercial broadcast and air traffic. The medium and high subbands are placed for maximum support of emergency personnel, given the capabilities of their equipment. Operationally, to reduce co-channel interference, emergency units are assigned separate uplink and downlink bands to the HOBS interoperability nodes but can communicate among each other using conventional push-to-talk Time-Domain Duplexing (TDD). The two lower UHF subbands are tunable to 1G and 2G allocations. 3G bandwidths of 20 MHz are supported but only for one CDMA overlay, traded off against the 2G capacity. The upper UHF band supports one PCS (personal communications system) band and one RF LAN band. The 11 GHz SHF band was chosen because it minimizes antenna size on the mast for van-to-van trunking at E1 or E2 rates.

The high-speed digital interconnect, in an illustrative implementation, would require three gigabyte per second buses to interconnect IF ADC streams to the two hundred digital Channel Isolation filters. These may be packaged in three shelves with the wideband ADCs. Medium-speed Raceway-class interconnect

Figure 8.   HOBS Hardware Block Diagram

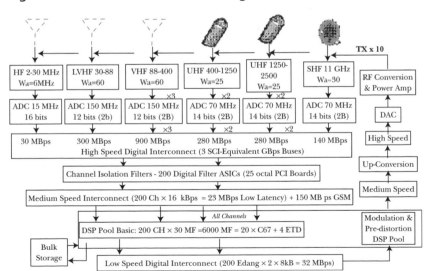

switches these signals to appropriate Texas Instruments (TI) C67-class DSPs or equivalent. Voice and data traffic requires 30 MFLOPS (MF) per Erlang or 6,000 MFLOPS. Assuming 60% efficiency of deliverable MFLOPS, ten C67 DSPs could provide this capacity. An operating margin of 50% spare capacity then requires twenty C67 chips. Since there are ten bulk streams, two chips (one dual C67 board with local and global memory) may be nominally associated with each bulk stream, for ten Erlangs of traffic per chip. These ten boards may be organized into one DSP pool shelf. For simplicity, the bus hosts are not shown. The DSP pool requires one shelf, and the bulk storage, LANs, hosts, etc. require an additional shelf. Alternative hardware implementations include the equivalent processing capacity in Power PC (PPC), or Intel Pentium/ Itanium processors in lieu of DSP chips.

The transmission facilities (DSP pool, upconversion, etc.) are sized by comparison to the receiver pool, 20% of the capacity of the receivers, or four C67 chips. The digital upconversion could be based on Intersil HSPs or Graychip GC4114 quad digital upconverter chips or equivalent. A shelf of eight octal boards provides sixty-four transmission channels switched to ten RF amplifier boards.

The system is configured into the van with a control rack (CTL) in the front, the receiving rack (RX) on the left side, and the transmission rack (TX) on the right side of the van. One operator position would be provided on the CTL rack for local technical control and mission planning. Five additional wireless laptops are packed for use near the van via RF LAN.

The back of the van has a swing-away auxiliary power unit liquid fuel-powered UPS to supply the substantial power requirements of the DSPs and RF transmission system. Antenna mast design minimizes EMI/RFI with physical separation, insulators, etc.

2) SDR Software Components. Software would be based on the SDR Forum SCA or the Object Management Group (OMG) equivalent Software Radio Architecture (SRA). This architecture specifies a POSIX compliant operating system with CORBA middleware functions, which may be implemented in Java. Applications software includes the following waveforms: AM, FM, and voice coder algorithms for single-channel voice, AMPS for 1G cellular, Digital

AMPS, IS-136, GSM, and 5 MHz W-CDMA for the cellular bands. There are TETRA, DECT, and PHS/PDC software packages for the DSP platforms. HF ALE, GSM Packet Radio System (GPRS), and V. modem software packages provide narrowband data connectivity. A Microsoft-based approach could include MS Office with Outlook providing word processing, database, and E-mail with Internet Explorer for web browsing. RF-CAD could be used for propagation prediction to site the vans and assist in managing spectrum allocations, with a Geospatial Information System (GIS) like ArcView. Spectrum policy software that might be based on XG could enable more flexible radio spectrum sharing in humanitarian operations as follows.

3. Ad Hoc Radio Spectrum Pooling in Humanitarian Operations.

HOBS mobile infrastructure offers opportunities for more flexible use of the radio spectrum in a disaster area. In such situations, greater flexibility in spectrum use may translate to more timely assistance dispensed more effectively. The basic idea is that those who currently own spectrum licenses could share radio spectrum with others using a real-time spectrum pooling etiquette, a polite asymmetrical protocol. The etiquette would permit spectrum sharing for discrete time intervals (minutes, hours, days, or more, as dictated by the situation). Enabled by HOBS, the etiquette would defer pooled spectrum back to legacy users, including RF-disadvantaged relief workers, within milliseconds. RF-disadvantaged relief workers are people whose conventional radios operate only on a hardware-defined set of frequencies, with a small number of waveforms/modes, and with little-to-no ability to upload AA software. Using etiquettes, licensed owners could donate radio spectrum for disaster relief. In addition, managers of humanitarian operations could delegate most of the details of spectrum sharing to their spectrum-pool-AA radios, saving time and improving relief team efficiency.

A. Spectrum Aggregation Strategy

Spectrum aggregation strategies implemented by policy-AA HOBS nodes would be those sanctioned by the local spectrum regulatory authority. The XG policy language could enable those authorities to rapidly adopt spectrum-use policies consistent with rapidly emerging needs in a disaster situation. In addition, XG policy-aware HOBS nodes could accumulate spectrum-use data, proposing policy changes to the local authority and further accelerating adjustments to spectrum-use policies as the needs of the humanitarian operation dictate. An aggregation strategy applicable to a disaster situation could be "If it is supplying or supporting emergency relief, give it spectrum; if not, then reduce its priority." Satellites and aircraft move rapidly and/or cover large areas, so the bands dedicated to these vehicles would not be pooled. Broadcast television stations, on the other hand, offer spectrum pooling and sharing opportunities enabled by recent regulatory decisions.[19] An effective spectrum aggregation strategy for humanitarian operations reflects the capabilities of mobile devices available at the scene along with the spectrum access capabilities of HOBS-class deployable infrastructure.

B. Mobile SDR Devices Enable More Agile Spectrum Sharing

In 1999, Mitsubishi and AT&T announced the first "four-mode handset." The T250 operated in TDMA mode on 850 or 1900 MHz, in first generation Analog Mobile Phone System (AMPS) mode on 850 MHz, and in Cellular Digital Packet Data (CDPD) mode. This was just the beginning of a proliferation of multiband, multimode, multimedia wireless products that now include 3G.

In the not-too-distant future, SDR PDAs could access satellite mobile services, cordless telephone, RF LAN, GSM, and 3G W-CDMA. Such devices could affordably transmit and receive in octave bands from .4 to .96 GHz, 1.3 to 2.5 GHz, and 2.5 to 5.9 GHz (avoid pooling the air navigation and GPS band from .96 to 1.2 GHz).

Not counting aircraft, satellite mobile, and radio navigation bands, such mobile radios would access more than thirty mobile subbands in 1463 MHz of potentially *sharable* outdoor mobile spectrum. The upper band provides another 1.07 GHz of potentially sharable indoor and RF LAN spectrum.[20]

C. HOBS-class Infrastructure Enables Opportunistic Spectrum Pooling

The laws of physics constrain spectrum pooling. The HF band, for example, propagates for thousands of miles, with kbps-class data rates. HF therefore provides HOBS with long-haul connectivity but is not suited for spectrum pooling within the disaster area. Bands above about 6 GHz (upper SHF) rely on directional antennas to obtain mbps data rates on LOS paths. Such highly directional paths enable the spatial sharing of spectrum, for example, by pointing the directional antennas on a HOBS mast. These bands are not suitable for forming spectrum pools of omnidirectional transmitters in which the relief worker need not care about pointing the antenna.

Table 3.   Mobile RF Spectrum Pools

Band	RF_{min}	RF_{max}	W_c	Remarks
Very Low	26.9	399.9	315.21	Long range vehicular traffic
Low	404	960	533.5	Cellular
Mid	1390	2483	930	PCS
High	2483	5900	1068.5	Indoor and RF LANs

The four spectrum pools of table 3, however, are ideally suited to mobile relief worker spectrum pooling. The *very low band* of this mobile spectrum regime penetrates buildings and propagates well in rugged terrain. The *low band* has the best propagation for high-speed terrestrial mobile traffic, in part because auto and rail traffic is supported with relatively low infrastructure density. *The mid band* is best for Personal Communications Services (PCS) with its higher infrastructure density. In addition, the *high band* has the large coherent bandwidth for high data rate Internet and mobile video teleconference applications. The 3G waveforms could be used in any of these bands but are best suited for the *low* and *mid bands*. W_c is the total spectrum that could participate in the spectrum pool based on an analysis of U.S., Canadian, and UK spectrum allocations. W_c does not include satellite, aircraft, radio navigation, astronomy, or amateur bands, which are not suited for pooling. An ad-hoc relief-worker spectrum-pooling concept enabled by HOBS mobile infrastructure is illustrated in Fig. 9.

Figure 9.   Spectrum Pooling by HOBS-class AA SDR

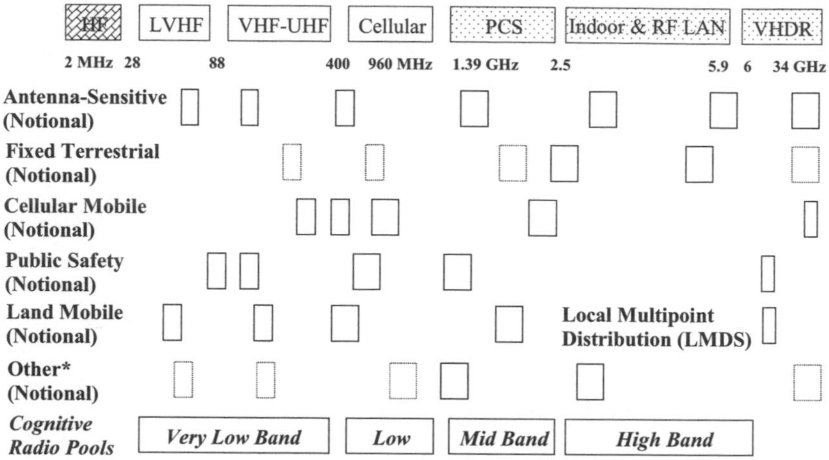

| HF | LVHF | VHF-UHF | Cellular | PCS | Indoor & RF LAN | VHDR |

| 2 MHz 28 | 88 | 400 960 MHz | 1.39 GHz 2.5 | 5.9 6 34 GHz |

Antenna-Sensitive (Notional)

Fixed Terrestrial (Notional)

Cellular Mobile (Notional)

Public Safety (Notional)

Land Mobile (Notional)   Local Multipoint Distribution (LMDS)

Other* (Notional)

Cognitive Radio Pools | *Very Low Band* | *Low* | *Mid Band* | *High Band* |

* Includes broadcast, TV, telemetry, Amateur, ISM; VHDR = Very High Data Rate

In a major disaster such as a category 5 hurricane, HOBS might provide gap-filling and emergency management in concert with a relatively large number of extant cell sites as illustrated in Table 4.

For simplicity, the entire population offers load (100% penetration). With pooling, each mobile outdoor user would have an average of 432 kbps. This assumes today's infrastructure density and 2G-equivalent bandwidth efficiency (0.2 mbps/MHz/cell). These rates are gross data rates not discounted for Quality of Service (QoS), which can lower these rates substantially if low bit error rates are required (for instance, for file transfers).[21] On the other hand, 3G technology achieves 0.45 mbps/MHz/cell, so the rates are representative of the range of rates achievable with a mix of 2G and 3G technology.

With ad-hoc spectrum pooling, then, multimedia bandwidths needed for extensive telemedicine support, for example, can be achieved without increasing the number of cell sites. The aggregation of public-use spectrum by SDR technology could transform police, fire, and other VHF-UHF towers into equivalent pooled-spectrum cell sites. Such SDR technology expands opportunities for the dynamic sharing of spectrum. But the well-heeled conformance to the radio etiquettes afforded by AA-radio is what could make such sharing practical.

Table 4.   Illustrative Pooled Spectrum Parameters in a Major Disaster

Parameter	Illustrative Range of Values	Remarks
Total Spectrum	0.4 to 2.5 GHz	1.463 GHz pooled
Duplexing	Frequency Domain (FDD)	Evolved from cellular services
Voice Channel	8 1/3 kHz-equivalent, TDMA or CDMA	Evolved from second generation
Channels per cell	25088	Usable, including 6:1 reuse and FDD
Coverage area	4,000 square kilometers	The size of e.g. Washington, DC
Number of cells	40 extant (plus 40 public sites*) plus HOBS	5.5 km average cell radius (3.9 km)
Population	609,000	The population of e.g. Washington, DC
Offered demand	0.1 Erlang	Multimedia level (vs. 0.02 for voice user)
Demand per cell	1,522 Erlangs	Drops to 761 considering public cell sites
Spectrum per user	160.7 kHz (320.4 kHz with public sites)	.22 to .64 Mbps/ user (.4 to 1.28 Mbps)

* Public sites are towers of police, fire, military, and other government/ public facilities pooling spectrum.

4. SPECTRUM POOLING ETIQUETTES FOR HUMANITARIAN OPERATIONS

Radio etiquette is the use of RF bands, air interfaces, protocols, and spatial and temporal patterns mediated via high-level rules of interaction that dynamically share pooled radio spectrum. Etiquette for spectrum pooling includes a spectrum renting process, a shared spectrum allocation process, assured back-off to authorized legacy radios, assured conformance to precedence criteria, and a policy coordination network overlay. For disaster relief, the policies must accommodate the rapid influx of RF-disadvantaged relief workers, as well as a mix of policy-AA handsets, vehicular radios, and HOBS-class mobile infrastructure. In the sequel, a regulatory authority operates a designated HOBS SUV, confirming or adjusting spectrum-use decisions made by pooling etiquette algorithms as follows:

A. Pooling or Renting Spectrum

A protocol framework for pooling or renting radio spectrum is illustrated in Fig. 10. This time line shows the power levels in the shared channel, differentiating signals of client and spectrum manager. The spectrum manager acting as offeror initiates the process by posting an "advertise" flag in the channel for rent. This in-band signaling accomplishes multiple goals. First, it unambiguously identifies the frequency, bandwidth (through its spectrum occupancy mask), and spatial extent of the channel (through the propagation of the signal). This signal would be pseudo-random, coded so that signal-processing gain recovers the signal when it is weaker than the noise and interference. A filtered PSK PN sequence of 100 bits duration at a 10 k chips per second rate, filtered to an 8 1/3 kHz bandwidth would advertise an 8 1/3 kHz channel in a 10 ms burst. The offeror listens for 10 ms and then repeats the offer-signal twice more. The sequence starts as close as practicable to the tick of the offeror's local GPS-second clock. The three-flag series repeats at the next GPS second.

A legacy user could hear these bursts in conventional FM channels, realize that the channel is available, and express a need to use it by keying the transmitter. In other words, one does not need an SDR or policy-AA radio in order to express a need to use a pooled radio channel. RF-disadvantaged users accomplish this simply by the act of using the channel. The HOBS-enabled spectrum management authority could verbally request the user to switch channels. With language-AA software, HOBS itself could automatically generate such a verbal request on the legacy channel by speech synthesis. Language-AA software could recognize the language of the reply and could in principle interact autonomously to determine the legacy radio's capabilities and to direct the user to an appropriate ad-hoc pool of legacy radios.

Figure 10.   Time Line of Spectrum Rental Protocol

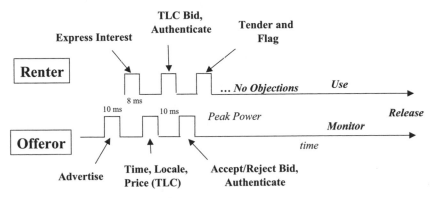

A rental-AA or pool-AA radio could express interest in the channel with a coded interest-burst similar to the advertise-burst. If the offeror hears the interest-burst, the second burst specifies the rental time interval, operating locale, and price of the channel, which would be zero in the case of cooperatively pooled or donated spectrum. This data exchange would be Huffman coded using a-priori knowledge. The renter then submits a bid with a short authentication sequence. The offeror may accept the bid, authenticating itself in return. Finally, the renter tenders the E-cash or spectrum barter s-cash, completing the initial rental protocol. Pool users simply accept the offer of shared spectrum to complete the transaction. HOBS pool-AA software tracks this ad-hoc reallocation of spectrum in its database to support both algorithmic planning and decisions of spectrum managers.

Returning to the etiquette, both parties to the sharing then wait and listen for the rest of this GPS second for objections. The 100-bit objection-sequence would be nearly orthogonal to the advertise- and interest-sequences. Any legacy radio that begins to use the channel in native mode (for instance, FM push-to-talk) automatically negates the rental agreement if its received signal strength at the renter or offeror location exceeds a threshold. A 100-bit rental-cancellation sequence from either party then cancels the ad-hoc pooling arrangement. The offeror cannot attempt to pool the channel again until after a specified waiting period (from seconds to minutes) or until the spectrum manager asserts that legacy users have been cleared from the channel. After using the channel successfully, a renter would provide additional E-cash or s-cash validation bits required to finalize the transaction. In the disaster-relief application, HOBS would accumulate statistics of such usage to support spectrum management decisions.

To avoid problems with the one-second granularity of the rental agreements, a service provider could provision the network by renting a few standby channels for traffic that cannot wait for the next rental period. During the use of the channel, both offeror and renter continue to use a polite back-off protocol as follows.

B. Polite Back-off Protocol: Defer to Authorized Legacy Users

If the pool-AA radios using the channel were to employ a conventional air interface, legacy users would be unable to break in. Thus, for example, a police officer could not use his previously assigned frequency to call for assistance. The polite back-off protocol illustrated in Fig. 11 solves this problem, albeit at the expense of some loss of throughput.

The renter supplies (digital) traffic to the channel for 20 ms as shown in the time line. Both renter and offeror listen during the subsequent 5 ms listen-window. The high rate of listen-windows assures that not more than 25 ms of legacy speech would be truncated, a level that should be essentially imperceptible distortion of a push-to-talk radio signal. If a legacy waveform exceeds a carrier to interference ratio (CIR) threshold, for either renter or offeror, the conflict is recognized and the channel is immediately

Figure 11.   Polite Back-off Protocol Assures Channel Access

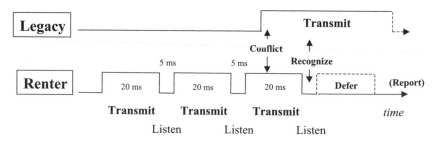

vacated. The truncating party issues a 10 ms termination-burst that indicates the cause is legacy interruption. The rest of the traffic would be moved to an alternate channel. Since E-mail, attachments, file transfers, audio clips, video clips, and other asynchronous multimedia are relatively insensitive to end-to-end delay, the movement to an alternate channel inserts a time delay that the higher levels of the protocol stack could accommodate.

5. Spectrum Policies for Humanitarian Operations

Spectrum pooling etiquettes enable spectrally efficient pooling of radio spectrum needed in humanitarian operations. In addition, higher-level polices are needed to enable greater operational efficiency of spectrum use. These policies may be formulated and expressed as a-priori XG policies, or they may be defined during the humanitarian operation and communicated using XG or an appropriate XML dialect.

A. Precedence and Priority

All users want guarantees that spectrum will be available when and where needed. Thus, any workable pooled-spectrum approach has to have a way of providing such guarantees. Fig. 12 provides an example of precedence of spectrum uses. Designated authorities may change precedence globally or locally. The etiquette allows one to designate a user (for example, by international mobile subscriber identification), a channel, or any combination of (user x time x space x frequency) with a specific priority for access or precedence in retaining use of a radio resource.

Those policies that are subjectively acceptable must be formal-

Figure 12. Precedence of Spectrum Use

1. **Emergencies**—*Established by authorities, inferred from events*
2. **Government**—*Attributed by band or channel modulation*
3. **Public Interest**—*Default by band, inferred from events*
4. **Commerce**—*Default by band and mode, inferred*
5. **Other**—*Recreational, sports, hobbies, etc.*

ized so that the radio control algorithms will perform as intended. In particular, the radios have to be able to infer many aspects of precedence from events. This is a technical challenge that requires the radios become more situation-aware. In the AA ontology of Fig. 3, enhanced situation-awareness implies greater ability to detect specific a-priori use cases, with proportionally greater ability to adapt to the situation. For example, a military radio generally may be considered to enjoy "government" priority, which is lower than "emergency" priority. In a disaster setting, however, military radios assigned to the relief operation could enjoy emergency status. A second level of use case is then needed to determine whether the military radios in use assisting fire fighting have priority over those assisting medical triage. Formalized exchanges of awareness knowledge enable more precise conformance to such use-case dependent details.

B. Formalized Exchanges of Awareness Knowledge

The spectrum manager would create a policy channel, an ad-hoc signaling and control channel in which policy-related knowledge may be shared from HOBS nodes to policy-AA subscriber radios. Policy-AA radios could create an ad-hoc policy channel using a peer network in which the first user becomes the policy manager. This is similar to the self-designation of network control stations (for instance, JTIDS).[22] The details of such a network are not central to AA and cognitive radio, but the language used to represent general world knowledge, plans, and needs is a key issue.

Some Radio Knowledge Representation Language (RKRL) is needed to formalize the depth of knowledge sharing needed to reliably detect humanitarian operations radio use cases. The Knowledge Query and Manipulation Language (KQML) was designed to facilitate the exchange of general knowledge.[23] Based on performatives such as "tell" and "ask," KQML can intuitively express pooled-spectrum management information. XML and its dialects are the more contemporary and more general purpose languages for expressing such knowledge. Although general purpose KQML-like tags in XML would suffice, the introduction of new RKRL-ontology referenced tags for spectrum pooling imparts additional structure to the dialogue. The new tags could include: Rental_offer, :RF_low, :Nchannels, :allowed_formats, :Legacy, :Equivalent, and tags representing standard PCS formats such as DECT, GSM, and 3G. The tags :From and :To refer to the time at which the rental is being offered. With a KQML/XML/RKRL construct, mobile nodes and networks may share plans about anticipated needs for spectrum so that it may be efficiently identified and rented. The formal plan to offer spectrum uses the tell performative to tell the AA network its plan to pool spectrum for a notional Hurricane Harold emergency as shown in Fig. 13. In this example, the radio also warns the network that its legacy users employ 25 kHz push-to-talk FM radios.

Figure 13.   KQML Expression of a Plan

{Tell :language RKRL :ontology Spectrum-Pooling:Emergency-Pooling
(:Event   Hurricane-Harold)
(:Spectrum manager Fairfax_Police :Location Chantilly_VA
(:RF_low 451 :Nchannels 12 :From 141118 :Until 141523)
:Allowed_formats ((:DECT 32kbps) (:GSM GPRS) (:Equivalent))
:Legacy 25kHzFM)}

The AA network uses an RKRL ontology to bind the semantics of the general knowledge expressed in the packet. This U.S. spectrum manager is willing to allow SDRs to employ a derivative of the Digital European Cordless Telephone (DECT) protocol and GSM GPRS in these groups of channels for this emergency. The other aspects of the plan to pool spectrum for the given emergency reflect the time, place, and RF channels the network plans to rent.

6. ENABLING DIGITAL TRIAGE

In October, UCSD announced WISARD, the Wireless Internet Information System for Medical Response in Disasters.[24] WISARD replaces felt pen and whiteboard with RF identification (RFID) tags for disaster triage. Some RFID tags measure vital signs when attached to a finger. Among the motivations cited by project director Leslie Lenert, M.D., was the recent scenario in which the Russian government used a gaseous agent to disable the terrorists in a Moscow theater. More than one hundred of their hostages died. According to Lenert, medical personnel later reported that most deaths were due to lack of vital signs monitoring at the scene and an inability to organize care to determine who was breathing and who wasn't. Immediate application of the RFID tags would have enabled digital triage, hopefully saving lives.

Figure 14.   WISARD Addresses Field Care Problems © UCSD, reprinted with permission

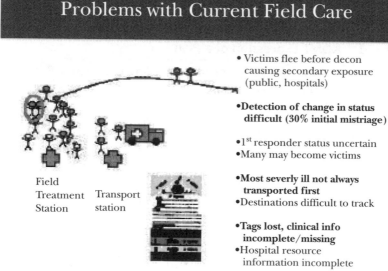

The project further envisions using IEEE 802.11 for the sharing of RFID data among PDAs and a central database. While IEEE 802.11 is in general a great idea, other communities of interest are also employing 802.11 for logistics coordination, air support de-confliction, and the like. If dozens of different humanitarian organizations arrive at a scene all using 802.11, the cacophony of the VHF bands will be inflicted on the 802.11 bands. HOBS and spectrum-AA radios need not suffer degradation of service in such cases, however. Groups of spectrum-AA radios on the scene would be directed to set up 802.11-equivalent networks in a pooled radio spectrum, such as in an unoccupied television broadcast band. In this situation, the entire 802.11 protocol, media access control, and even physical layer frequency hopping is moved from its normal 5 GHz carrier to, say, an unused band centered at 700 MHz. Although radio propagation would be substantially different in this low band, the 802.11 protocol can accommodate the differences. Thus, pooling and agility of spectrum-AA radios, enabled and managed by HOBS nodes, dramatically reduces spectrum crowding in the over-used 802.11 standard band.

Emerging RKRL and humanitarian operations ontologies and XML/XG/KQML-class knowledge exchanges enable spectrum-AA algorithms to interrogate arriving 802.11 network applications for intended use in support of the operation. Detailed use cases would prescribe which types of intended use have priority. In some cases, hardware-defined radios may not be able to move the 802.11 waveform to an alternate RF, so the HOBS spectrum planning software would examine alternative plans and could propose a spectrum-use solution that maximizes value with respect to the criteria of the use cases. Without such ontology-capable radios, people are left to negotiate these details on the scene, potentially wasting precious time better used for digital triage than for spectrum de-confliction.

7. The Complexity of a Society of Cognitive Radios

Given a spectrum pooling framework as outlined above, spectrum-AA radios could automate much of the spectrum management that now is done manually. One might think that the performance of cognitive radios that are AA and that also learn from experience could be projected using contemporary radio-engineering techniques. But cognitive radios' learning ability endows them with a complex internal structure, changing behavior in response to local circumstances. No two CRs will have exactly the same experiences. Since CRs also will be richly interconnected, they will form a complex adaptive system.[25] CRs therefore might behave like an ant colony, evolving their own paths through the spectrum and intervening nodes to ferry voice, data, video, and multimedia packets through the ether.[26] These radio-ants might move packets from the smaller power-starved radios through multimode vehicular radios and on to conventional cell sites. If the etiquette is too strict, very little additional benefit will come from cognition (that is, from the machine learning) over the AA spectrum pooling because there will be insufficient degrees of freedom to overcome local maxima in the learning search space. If, on the other hand, the etiquette is too liberal, there may be much unintended interference and potentially universally poor quality of service. Such complex adaptive systems operate best "at the edge of chaos."

The edge of chaos is not a particularly comfortable place for spectrum managers. Such complex adaptive systems are in fact difficult to understand, model, and diagnose.[27] Nevertheless, they also produce efficient answers to NP-hard problems. Thus, in some sense an ant colony of cognitive radios left to evolve spectrum use among themselves might be the most efficient way to achieve high value from limited radio spectrum in a reasonable time. How can we structure the capabilities and etiquettes of cognitive radio so that spectrum pooling is workable? There are many important aspects to this question. Global CR research continues to address such questions.

8. Conclusion

SDRs, AA radios, and ultimately CRs provide a vast untapped potential to adapt radio resources and wireless services to the unique needs of humanitarian operations. The contemporary process of spectrum allocations takes years and lacks flexibility but provides necessary access guarantees and accountability for spectrum abuse. In part, this is because the technology for guaranteeing spectrum use according to policy to primary spectrum managers is in its infancy. SDR, AA, and ultimately CR offers opportunities to advance spectrum pooling technologies for better sensitivity to users' needs in a given situation. HOBS, AA radio, and the enabled spectrum pool etiquettes are thus offered as an approach to more efficiently use a limited resource that is in high demand. Agent knowledge, computational ontologies, and related reliable inference mechanisms are developing through the semantic web community. The emerging generation of increasingly intelligent radios therefore offers great promise for enhanced commercial, civil, military, and humanitarian operations.

ACKNOWLEDGEMENT

The author would like to thank Drs. Preston Marshall (XG) and Larry Jackal (cognitive radio) of DARPA, Mitch Kokar of Northeastern University (ontology-based radio), and Cliff Weinstien of MIT Lincoln Laboratory (natural language processing) for continuing stimulating technical dialogues on the broad range of topics represented here.

In simulated "conflict negotiation" exercises the participants are often pleased with their results and the speed at which they achieved them. They are a little disappointed when they realize that in reality one usually must at least double the time to allow for adequate interpretation. Even the finest simultaneous interpreters will add up to a third of the time.

In the earliest hours of an emergency, the response team will have to rely on whomever it finds for interpretation. The conversations may be limited in vocabulary but will be extensive in results. Where is your leader, may I see him? Where are those who are likely to be shooting at us? Are there minefields? What is the priority for aid? Is it food or medicine or shelter? If I bring in an aircraft do you have equipment to unload it? Where can I hire trucks? Where are the refugees? Can I get to them? These conversations may be with farm laborers, local leaders, ministers, or even the president.

There will be the need for agreements to be drawn up, contracts to be exchanged, and documents to be read and written. Early translations can have enormous long-term effects; early poor translations can have devastating and binding effects.

At the personal and most satisfying end of humanitarian work there are the personal contacts, the new friends made, the new heroes to be admired.

Most medical aid workers can tell stories, romantic in the telling, and even more so in the retelling, of being awakened in the night and asked to follow a distressed father to a local house to visit a sick wife or child and then attempting to diagnose by candlelight and sign language. The reputation of the agency and the credibility of the medical team may well be judged by the outcome of the visit.

Before long the smell of money will bring potential interpreters to the office location. The best of them are usually from the professions and have studied overseas. But is it right to take a doctor from the hospital, an engineer from a power station, or a teacher from a school to translate at checkpoints?

As the crisis settles down interpreters become an essential part of the team: the mouthpieces of the operation. This gives them power and money, but it also attracts great risks. They are identified with the decisions they interpret, most are threatened, and some are abused. When events settle down and the emergency teams leave, the interpreter is left to merge back into the community, sometimes with fatal results. Computer-aided language translations could offer rapid, safe, and significant help.

—Larry Hollingworth

needs such as emergency famine relief, coping with a sudden-onset epidemic, or accompanying a peacekeeping mission do not allow for a typical six-month or one-year language education "crash-program." Another potential solution is to ferry a suite of human translators along with the aid workers. That too is fraught with inadequacies, including high expense, exposure of additional personnel to the local dangers (disease, insurgency, etc.), and the sheer difficulty of finding translators for minority languages, let alone enticing them to participate. Therefore, in actual practice, a makeshift combination of approaches is followed, including on-the-job rudimentary language learning, occasionally finding willing translators, and simply going without—a very risky proposition.

This chapter explores a technological solution to the minority-language communication challenge—or at least an important technological ingredient to a combined solution augmenting scarce human translators, if available. That solution entails combining multilingual speech recognition and speech synthesis with new machine translation technologies. Recent advances in all three areas hold significant promise with respect to producing acceptable levels of accuracy, especially for targeted domains (for example, medical interviews). Moreover, at Carnegie Mellon University, in conjunction with our partner spin-off companies, we have integrated and miniaturized the three technologies to produce increasingly functional early prototypes of handheld speech-to-speech translation devices. These devices operate in three phases:

1. Recognize the spoken language in one language—let us call it the source language—optionally confirming the corresponding text with the speaker to correct potential errors in the speech recognition.
2. Translate the source language text into the second language—the target language—optionally translating back to the source in order to detect and correct potential translation errors.
3. Synthesize the translated text into speech in the target language.

If the source-language speaker is illiterate, or if the speaker has gained sufficient confidence about the system's ability to recog-

Language Technologies for Humanitarian Aid

Jaime G. Carbonell, Ph.D., Alon Lavie, Ph.D., Lori Levin, Ph.D., and Alan W. Black, Ph.D.

INTRODUCTION: THE NEED FOR SPEECH-TO-SPEECH TRANSLATION

Humanitarian aid missions, whether emergency famine relief, establishment of medical clinics, or missions in conjunction with peacekeeping operations, require on-demand communication with the indigenous population. If such operations take place in countries with a commonly-spoken major language, such as English or Spanish, it proves relatively easy to find participating personnel with the appropriate linguistic fluency. However, such is not the case when the operations take place in regions where less common languages are spoken, such as Bosnia (language: Serbo-Croatian), Haiti (language: Haitian Creole), Somalia (language: Somali, a.k.a. "Soomaaliga"), or Afghanistan (language: Pashto, with subpopulations of Urdu and Tadjik speakers). Even in Latin America, where Spanish and Portuguese dominate, there are more than one hundred indigenous languages, including Quechua in Peru, Aymara in Bolivia, Mapudungun in southern Chile, and the Tucan languages in the southern Colombian Putumayo region. Many native speakers of these languages are not versant in either Spanish or Portuguese, especially those in remote mountainous or jungle regions, where the need for medical or educational aid, or protection from organized drug gangs, may be paramount.

The "obvious" solution is to educate relief personnel in the native language of current interest, in order to reach at least a rudimentary level of communicative fluency. However, such education requires time and expense and must be repeated with all new personnel before they are rotated in. Moreover, sudden nize speech and translate, the confirmatory steps are omitted, and the translation proceeds faster, albeit with a potential for undetected errors.

Applying off-the-shelf technology for speech recognition and machine translation proves insufficient for the task. Speech systems are deeply customized to a given language, such as English, and are not easily adaptable for minority languages. Developing standard machine translation technology for a new language-pair (for example, English to Arabic) requires person-decades of specialized computational linguists and thus puts the effort beyond the economic reach of humanitarian aid applications. Instead, new technologies aimed at rapid low-cost adaptation to new languages are required, and those are precisely the ones under current investigation, as discussed in this chapter.

The remainder of the chapter is organized into three sections:

- Speech Recognition and Synthesis (corresponding to steps 1 and 3 above), including advances permitting rapid adaptation to new language.
- Machine Translation Technologies (corresponding to step 2 above), including the AVENUE project for rapid creation of translation systems for minority and endangered languages.
- Speech-to-Speech Translation (integrating all three steps), including discussion of a progression of projects: JANUS, DIPLO-MAT, Speechalator, with increasing capabilities and discussion of future prospects and applications in humanitarian aid, bilingual education, and preservation of endangered languages.

SPEECH RECOGNITION AND SYNTHESIS

Speech is the most natural form of communication for people; however it is far from the easiest form of communication for machines. Over the past thirty-to-forty years the processing of human speech by computer has advanced to the stage that it can now be used effectively in many practical situations. *Automatic speech recognition* is the process of converting audio recordings of human speech into text; *text-to-speech synthesis* is the inverse process of converting text into spoken, fluent audio. Both processes present their own major technical challenges, which we review in this section.

For automatic speech recognition, we must statistically model the acoustic variations that speakers use in speaking their language, as well as filter background noise. Phonemes, the fundamental units of speech, may be spoken in different ways depending on the other phonemes around them. The process is called "co-articulation." For example, the pronunciation of the consonant /s/ is acoustically distinct, although similar, depending on the shape of the following vowel. In a word like *so* the lips are rounded for the /s/, while for *see* the lips are not. Although human ears have learned to deal with such variation without even noticing its existence, automatic computer speech recognition needs to model these variations explicitly in order to recognize every appropriate form of every phoneme in context. Thus, the first step in building sufficient models is to collect examples of such speech in as many contexts as there are variations.

There are other levels of variation too that must be covered. Female and male speech are different, and children's speech also differs due to the size and maturity of the vocal tract. People also speak in different styles. Casual or slurred speech and precise speech are quite different; a speech recognition system must be able to handle all forms. Environmental conditions also cause variation. People speak differently when outdoors versus in a quiet office or on the phone. Linguistic factors also affect speech; a person's dialect, education, and social position can affect pronunciation. Human listeners are good at adapting to differences in human speech even when heavily accented. Speech recognition engines must often also deal with nonnative speakers, both with subtle and strong accents.

Speech output, on the other hand, should be clear and consistent, and may sometimes be based on a single speaker. Issues such as gender of voice and style of voice can, however, be important and may require a small handful of "voices." For instance, a command voice is needed when issuing an order such as "Put down your weapon now!" A more compassionate voice is appropriate when saying, "We are here to help" or asking, "Where does it hurt when you walk?"

The Phoneme Level

The first basic task in building speech models for new languages is the definition of a *phoneme set* for the target language. Phonemes are the fundamental pieces of speech that make up a language, such as the pronunciation of individual letters. The linguistic definition of a phoneme is a unit that when changed can lead to a new word. For example, the /p/ in the English word *pat* is a phoneme, since if it were to change to /b/ we would get the new word *bat*. The International Phonetic Association (IPA 1993) has gone far to define a set of phonemes that cover most of the variations of the languages of the world. But there are still subtle questions that often need to be addressed; even in major dialects of English, which are very well studied, there are questions about how many phonetic distinctions should be made. For example, in British English, the words *Mary, marry,* and *merry* all have different vowels, whereas in American English, for many people, there is no reliable distinction in their pronunciation.

The Lexical Level

Once a phoneme set is defined, the next stage is to construct a lexicon to map from words to sequences of phonemes. For some languages, such as Spanish, where the written form is close to the pronunciation, this phonology-orthography mapping can be done by simple rules. But for other languages with more complex relationships between orthography and pronunciation, such as English and French, a lexicon is required with explicit entries. Even the largest list of words in a language will never be 100% complete. Proper nouns, words borrowed from other languages (for instance, *sushi, au contraire, ombudsman, macho, insallah*), and neologisms like *gigabyte* or *defibrillator* will always pose new challenges to the most complete of lexicons. Thus, for speech recognition and synthesis we also need to be able to generate the most plausible pronunciation for an unknown word, just like humans do. This we can do by building statistically trained letter-to-sound rules from the lexicon. We have used a simple but reliable technique for doing this for a number of languages (Black et al. 1998).

In order to construct the basic lexicon itself we have developed a bootstrap technique (Maskey et al. 2004). In this technique we first hand-specify the pronunciations of some three hundred common words and then construct a set of letter-to-sound rules automatically from these entries. Then, using text in the target language, we find the most frequent words and test them against this letter-to-sound rule model. If they are correct, we add them to the base lexicon, and, if wrong, we hand-correct them, add them to the lexicon, and retrain the rules with the additional verified data. By iterating this technique we can quickly construct reliable lexicons even for languages with more opaque ortho-graphic to phonetic relationships.

Acoustic Recognition

In order to build speech recognition acoustic models we must have examples of speech in as wide a variation as possible within the intended use and subject matter of the recognizer (e.g., medi-cal interviews). Traditionally, speech recognition acoustic models require about one hundred hours of recorded and transcribed speech for training. It is crucial that the transcription reflex ex-actly what was actually said (including repetitions, false starts, etc. that are common in even quite careful speech), rather than an idealized or cleaned-up version, or else the acoustic training will fail to find the sound-to-text correspondences reliably.

The GlobalPhone Project (Schultz, 2002) has reduced the amount of data required to train a speech recognizer in a new language by using initial models from other languages and then adapting those models for the target language using a much smaller amount of data (Schultz and Wiabel 2001). The Global-Phone Project offers not just the ability to build new recognizers in new languages but includes a data collection component that defines and provides tools for non-speech-scientists to collect tar-get language data. GlobalPhone has already collected data from fourteen languages and continues to collect data for new lan-guages. This data repository also aids in moving to new languages by growing the common set of cross-language phones and build-ing clusters of related languages.

Although the best results can be achieved by increasing amounts of data from the target language, comparable results can be achieved with relatively small amounts of target language data complementing data available from other languages. Using initial multilingual models plus as little as around one hour of transcribed target speech results can achieve results similar to that of collecting tens of hours of speech in that language. This is very important for rapid development of speech-based systems for new languages in emergency-aid situations, as the process of exact acoustic transcription is very slow and detailed—it takes ten to twenty hours to transcribe exactly one hour of speech.

Text to Speech Synthesis

For speech synthesis, unlike speech recognition, we can often limit our scope to a single voice, or perhaps just one male and one female voice. The FestVox Project (Black and Lenzo 2000a) offers tools, techniques, and documentation on how to build synthetic voices in new languages reliably, without requiring a computational speech scientist. The technological approach is termed *concatenative speech synthesis.* Appropriate small sub-word units of natural speech are selected and concatenated to form words and new utterances. The quality of the process can approach that of recorded human speech, though unlike recorded speech, it can be used to say unanticipated words, phrases, and sentences, for instance those produced by a machine translation system. The design of the recorded database is, however, crucial; it must cover the phonetic and prosodic space.

The data collection process proceeds as follows: We first collect a large amount of text in the target language. Then we select short sentences (say, fewer than twenty words) that contain primarily high frequency words. Such sentences are typically easy to say; pronunciation errors are thus minimized. We then use the constructed lexicon to convert the text into phoneme strings, as discussed above, and focus our pronunciation training on the sentences with the best phonetic coverage. We repeatedly select sets of phonetically rich sentences until we have identified around one thousand such sentences. In addition to general text we may also include targeted text for the particular application, such as medical interviews. The closer this designed database is to the target utterances, the better the quality of the synthesis. In extreme cases we can design the system to cover a targeted domain (Black and Lenzo 2000b), augmented with standard greetings and transition phrases. The database is recorded by a single native speaker of the target language in a studio-quality environment. The data is then automatically labeled using a speaker specific acoustic model to find where all the phoneme boundaries are. For best results these labels are hand corrected, but that is a resource intensive task.

Evaluation of the speech output is extremely important. Just because it may sound Chinese, Greek, or Quechua to the builder of the voice does not mean it sounds natural to native speakers. Evaluations including listening tests by natives are used to ensure that the quality is acceptable and understandable. A number of different tests explicitly measure phonetic coverage and domain coverage (Tomokiyo 2003).

Challenges for Minority Languages

As we cover more languages, our tools and techniques improve. But from this wider coverage, we also learn that there are other factors that can make the construction of speech technology in new languages harder. The top world languages have substantial amounts of written text, linguistic analysis, and large volumes of text readily available on-line. As we move to the less-spoken languages, such resources become more and more scarce. Phonetic systems for minority languages may not be defined, lexicons may not be readily available, and written texts may be available only on printed or handwritten media.

Whereas the major languages of the world have standardized both their orthographic and phonetic conventions (spelling and pronunciation), the same is not true for many minority languages. For instance, Mapungun, spoken in southern Chile and Argentina by the indigenous Mapuche population, has several distinct orthographic variants and at least an equal number of phonological ones. Even for majority languages, standardization may be partial or recent. For example, although a well-defined version of Arabic exists, Modern Standard Arabic, this is not normally a spoken language. The people in daily conversation use their own dialects, differing in both pronunciation and lexicon. Building a speech recognizer and synthesizer in Arabic requires first a decision about which dialect(s) to choose. Then, once chosen, we must ensure that we build lexicons and orthography-to-phonetic mappings for that dialect rather than simply Modern Standard, even though web-available material is far more abundant for the latter.

There are also sociolinguistic issues in building speech and language models. For instance, many cultures are more gender sensitive than English-speaking ones. The grammar and marking within the language may change depending on the gender of the speaker, not just that of the addressee. Such language models need to be added to the system so that a female synthesized voice uses appropriate female language, while the male output uses appropriate male terms. Such gender issues can be especially important in dealing with sensitive subjects such as medical interviews that may refer to anatomical concepts, hygiene, diet, family, or reproduction.

Another interesting issue is whether the speech-translation system should produce synthetic speech that sounds like a native speaker or has an accent typical of the source language speaker (for example, American). In the development of a Pashtu synthetic voice, we used a U.S. English speaker, trained in phonetics, to mimic the natural Pashtu speech, as no native Pashtu speaker was available. Thus the resulting synthesizer had a slight American accent and to Pashtun natives sounded nonnative, which, surprisingly, they thought was just perfect, as they could see that the original speaker was American and so his translated voice was appropriate and not deceptive.

Machine Translation Technologies

Machine Translation (MT) (Hutchins and Somers 1992) has become a popular technology on the Internet. Many websites offer free automatic translation, and some search engines, such as Google, offer to "translate this page" automatically if the language of the page is different from the language of the user's web browser interface. However, closer inspection reveals that MT is available for very few language pairs. A language pair consists of a source language—the language one is translating from—and a target language, the language one is translating into. Free web-based translation services are generally available for pairs of major

European languages (usually English, Spanish, French, German, Italian, Portuguese, and maybe Russian) and for a few pairs of European and Asian languages (usually Japanese and Chinese). The Compendium of Translation software (http://www.eamt.-org/compendium.html) lists all MT software available for sale. In this list we can find more language pairs. However, with a few exceptions, most of the source and target languages are spoken in countries that can provide a large consumer base for MT systems.

Many of the commercial systems available today are the results of person-decades of work and therefore are developed for language pairs where economic prospects are favorable. Unfortunately, such economic imperatives exclude most minority languages where MT is most needed for humanitarian purposes. For this reason, there is a growing amount of research on producing MT systems for new language pairs quickly and cheaply. After a brief discussion of the state of the art in MT, we present our CMU AVENUE system for building cost-effective MT of minor languages.

Why Is MT Hard?

A naive concept of translation might involve looking up each word in a dictionary to find an equivalent word in the target language. Consider the translation of a simple sentence like "Me llamo Maria" into English. A word-for-word translation would result in "myself call Maria," which is not a good English sentence. A slightly more sophisticated word-for-word translation might be "Myself I call Maria," taking into account that "llamo" means "I call." Notice that this involves knowledge of morphology, the combinations of prefixes, suffixes, and root words, and that one word in Spanish can correspond to more than one word in English. However, we are still far from the correct translation, "My name is Maria." A major problem is that even closely related languages do not use the same word order. Another problem is lexical ambiguity (for instance, "llamar" means both "to call" and to "be named," the latter applying to the reflexive construction).

Semantics, the meanings of words and sentences, must also be taken into account. Perhaps the oldest MT joke involves the translation of "The spririt is willing but the flesh is weak" into Russian and back into English as "The wine is good but the meat is rotten." How can we know whether the word "spirit" refers to a soul or to an alcoholic beverage? A less poetic example involves translating "lock" into Spanish. As a verb, its meaning must be decomposed into "cerrar con llave" (literally, close with key), because there is no corresponding lexical item. These problems are more severe for distant language pairs.

In short, MT is hard because it involves knowledge about morphology, syntax, and semantics for both the target and source languages. Native speakers of a language have this knowledge in implicit form, but coding the knowledge explicitly in computer programs is a deceptively complex task because of the myriad subtleties of language.

What Can Be Done?

The intractability of the translation problem is usually addressed at a practical level, for instance, by specializing for the kind of task for which an MT system is being developed.

If the translations are going to be used for dissemination of information, such as technical instructions or public health alerts, then the output must be of very high quality so as to be understood accurately. There are several options in this case: (1) The MT system could interact with a human translator, who can catch and correct system errors, thus automation is partial, reducing overall human error, or (2) The semantic domain can be limited so that words do not have so many meanings (for example, in texts about food, "spirits" would only refer to beverages), or (3) the input can be restricted to an unambiguous subset of the source language. Different MT systems may be used for assimilation of information, for example, scanning news wire for information about outbreaks of infectious diseases. In this case, the users do not have control over the input to the system, but errors can be tolerated as long as the general idea gets across.

MT systems can also be used for dialogue, and these are the ones of central interest in this chapter. MT systems for spoken language face the additional problem of starting with the imperfect output of a speech recognizer. This tends to reduce the quality of translation. However, since two humans are participating in the conversation, they may be able to detect misunderstandings, correct them, and adjust their language or speaking style to reduce future errors. Speech translation systems therefore do not require full accuracy, but because of the difficulty of the problem, they are usually constrained to a limited semantic domain, such as search-and-rescue situations, primary medical care interviews, or negotiating travel arrangements.

As a practical problem, MT system development must also take into account the language resources that are available in the source and target languages. Resources can be manifold: text in electronic form, bilingual dictionaries, grammar rule sets, and/ or linguists who can write rules for morphology, syntax, and semantics. Translation rules written by linguists require substantial time and effort to debug, especially as the number of rules increases and it becomes difficult for rule writers to keep track of the behavior of each rule and the interactions among different rules. It may also be the case that linguists trained in computational transfer-rule writing are not available for some languages.

The dominant MT paradigm is rule-based, but translations can also be calculated automatically from parallel corpora—large volumes of humanly translated text such as the collected proceedings of the United Nations (Brown et al. 1990; Vogel et al. 2000; Hutchinson et al. 2003). Various algorithms can be used to calculate a translation model—a set of probabilities for translating source language words and phrases into their target language equivalents, augmented with target language models that prefer certain word sequences over others. For instance, "pichi wentru" in Mapudungun could translate into "young man" or "diminutive man," but the former is preferred as being a word sequence more typically used in English. This approach is called Statistical MT. Alternatively, Example-Based MT systems compare incoming source text to sentences that are in the parallel corpus. If parts of the new source sentence are similar to parts of previously translated sentences in the parallel corpus, the corresponding translations of the similar parts are assembled into a candidate target sentence. Example-based MT also uses a target language model to select which of potentially many alternative translations is the most probable in the target language.

Corpus-oriented algorithms have the advantage of not requiring human rule writers and thus can be brought on-line very quickly, given a large bilingual parallel corpus. But they have some disadvantages, the largest being that large human-translated parallel corpora simply do not exist for minority languages. Smaller corpora produce less accurate translations, and even these seldom exist in sufficient quantity to build meaningful statistical models for minority languages.

The AVENUE Project

AVENUE is a research project aimed at finding quick, low-cost methods for developing MT systems for minority or endangered languages, especially languages that do not have enough resources for corpus-oriented approaches to MT. We also aim to circumvent the large cost in time and money of manually writing a comprehensive set of translation rules. Therefore, AVENUE learns from corpora but from extremely small, linguistically balanced ones, and instead of learning probabilities, it learns translation rules that can be examined and extended by human linguists if there is a human linguist with adequate training in MT and fluency in source and target languages. Information about morphology and syntax are implicit in the probabilities that are learned by data-oriented methods but explicit in the rules learned by AVENUE.

AVENUE operates in four stages: learning, translation, decoding, and refinement as shown in Fig. 1. The details of these four stages are described in Probst et al. (2002) and Lavie et al. (2003). Here we present an overview of the first stage, learning, which has two sub-phases. Development of an MT system for a new language pair starts with the process of elicitation, which produces the data needed for automatic rule learning. Elicitation requires a user who is bilingual in the source and target languages but does not need to know linguistics and does not need to know how to write rules for an MT system. The elicitation interface is shown in fig. 2. A sentence is presented to the bilingual informant, who then translates it and aligns the corresponding words between the original sentence and its translation. The informant simply clicks on a word in each language, and the elicitation interface draws a line connecting them. The elicitation interface also produces an

Figure 1.   The Architecture of the AVENUE Framework

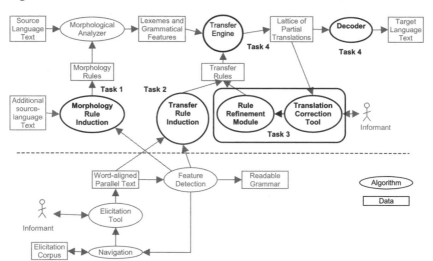

Figure 2.   The AVENUE Elicitation Interface

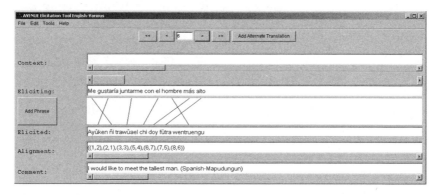

internal representation of the alignments in the form of indices such as (1 1) (first word aligns to first word), (3,5 2) (third and fifth words align to second word), and so on. Words may align with phrases (for instance, "lock" with "cerrar con llave") or with multiple disjointed words (for example, "not" with "ne" and "pas"—even the latter two are not always adjacent in French).

The elicitation interface has been used by Mapudungun speakers from Chile translating from Spanish, Aymara speakers from

Bolivia translating from Spanish, Hindi speakers translating from English, and Hebrew speakers translating from English.

The output of elicitation is a small but very useful parallel corpus of a few thousand sentences whose source and target words are carefully aligned. This corpus is the input to the rule learning component. Each portion of the corpus consists of *minimal pairs,* pairs of sentences that differ in only one fundamental linguistic way, such as singular-plural (to elicit pluralization rules) or a noun phrase with and without adjectives to determine whether phrase structures are head-initial or head-final (that is, whether the adjectives come before or after the main noun in the minority language). The rules learned can be fairly complex, as illustrated below.

Figure 3 shows an example of a translation rule for Chinese

Figure 3. An Example Translation Rule

```
; Rule to transfer Chinese question sentences
  {S,3} ; Unique rule identifier
  ; production rules: SL and TL type and constituent or POS
  sequences
  S::S : [NP VP "_"] -> [AUX NP VP]
  (
  ; Constituent alignments
(x1::y2) ; NP to NP
(x2::y3) ; VP to VP
; Parsing (x-side) constraints, build feature structure
((x0 subj) = x1) ; Assign NP's features to subj
((x0 subj case) = nom)
  ((x0 act) = quest)
    (x0 = x2)
    ; Transfer (xy) constraints
    ((y2 case) = (x0 subj case))
    ; Generation (y-side) constraints
    ; Insert AUX on target side based on
    ; value constraints
    ((y1 form) = do)
    ; Enforce value and agreement restrictions on y-side
    ((y3 vform) =c inf) ; verb must be infinitive
    ((y1 agr) = (y2 agr))
```

and English. In Chinese, in order to form a question that requests a yes-or-no answer, a question word is inserted at the end of the sentence. In English, on the other hand, a yes-no question begins with an auxiliary verb, as in "Do the children eat pizza?" In the rule, x0 refers to the Chinese sentence, x1 refers to a noun phrase that is the first element of the Chinese sentence, x2 refers to a verb phrase that is the second element of the Chinese sentence, and x3 refers to the question word. Similarly, y0 refers to the English sentence, y1 refers to an auxiliary verb such as *do* that is the first element of the English sentence, y2 refers to a noun phrase such as *the children* that is the second element of the English sentence, and y3 refers to a verb phrase such as *eat pizza* which is the third element of the English sentence. The rule shows that x1 should be translated into y2 and that x2 should be translated into y3. It also contains various constraints on the source and target language syntax. For example, y3 must contain an infinitive verb such as *eat* rather than a past tense or participial verb such as *ate, eaten,* or *eating.*

Rules like the one in Fig. 3 can be written by a human linguist or can be learned automatically from the output of the elicitation process. In order to learn rules automatically, the AVENUE system must capture two properties of human language syntax, compositionality and generality. Compositionality refers to the composition of larger phrases from smaller ones. For example, a sentence is made from a combination of noun phrases, verb phrases, prepositional phrases, and adverbs. A noun phrase can be made from adjective phrases, articles, nouns, prepositional phrases, and possibly also embedded sentences. The sentence in "Yesterday very big trucks brought the sacks of grain that were needed" could be described as two adverbs, *yesterday* and *very,* an adjective, a noun, a verb, an article, a noun, a preposition (*of*), a noun, a relative clause marker, an auxiliary verb, and a passive verb. However, it would be better to describe it as an adverb followed by a noun phrase, a verb, and another noun phrase, as shown in Fig. 4 below. The reason is that the latter description can also be used to describe other sentences that are similar but not identical in structure, such as *Usually, excess rain ruins crops* or *unfortunately the truck hit a pothole.* The AVENUE rule learner must therefore be able to recognize which parts of a sentence can be

Figure 4. The Compositional Phrase Structure of an English Sentence

Yesterday	adverb
noun phrase	
very big	adjective phrase
trucks	noun
brought	verb
noun phrase	
the	article
sacks	noun
of grain	prepositional phrase
that were needed	embedded sentence (relative clause)

grouped together into noun phrases and prepositional phrases. Then it must be able to hypothesize rules that compose those phrases into sentences. In order to accomplish this in a new language, the phrases of the source language (English or Spanish) are used as a guideline. The words that are aligned to words of the English or Spanish noun phrase are assumed to form a noun phrase in the new language as well. This is not always accurate, but it is usually a good starting point.

This compositional analysis permits us to induce translation rules, such as the example in Fig. 3, via a machine learning method called *seeded version space learning,* which is beyond the scope of this chapter.

DOMAIN-LIMITED INTERLINGUA-BASED SPEECH TRANSLATION

Evolution of Domain-Limited Interlingua-Based MT at CMU

The Language Technologies Institute together with the Interactive Systems Laboratory at Carnegie Mellon have been pursuing an ongoing research effort over the past fifteen years to develop machine translation systems specifically suited for spoken dialogue. The JANUS-I system (Woszczyna et al. 1993) was developed at Carnegie Mellon and the University of Karlsruhe in conjunction with Siemens in Germany and ATR in Japan. JANUS-I translated well-formed read speech in the conference registration domain with a vocabulary of five hundred words. Advances in speech recognition and robust parsing over the past ten years then enabled corresponding advances in spoken language translation. The JANUS-II translation system, taking advantage of advances in robust parsing (Lavie 1996), operated on the spontaneous scheduling task (SST)—spontaneous conversational speech involving two people scheduling a meeting with a vocabulary of three thousand words or more. JANUS-II was developed within the framework of an international consortium of six research groups in Europe, Asia, and the United States known as C-STAR (http://www.cstar.org). A multinational public demonstration of the system capabilities was conducted in July 1999. More recently, the JANUS-III system made significant progress in large vocabulary continuous speech recognition (Woszczyna 1998) and significantly expanded the domain of coverage of the translation system to spontaneous travel planning dialogues (Levin et al. 2000), involving vocabularies of more than five thousand words. The NESPOLE! System (Lavie et al. 2001) further extended these capabilities to speech communication over the Internet and developed new trainable methods for language analysis that are easier to port to new domains of interest. These were demonstrated via a prototype speech-translation system developed for the medical assistance domain called the Speechalator (Waibel et al. 2003).

Overview of the Interlingua-Based Approach

Throughout its evolution over the course of more than fifteen years, the speech translation systems have established a framework based on a common, language-independent representation of meaning, known within the MT community as an *interlingua*. Interlingua-based machine translation is convenient when more than two languages are involved because it does not require each language to be connected by a set of translation rules to the other in both directions. Adding a new language that has all-ways translation with existing languages requires only writing one *analyzer* that maps utterances into the interlingua and one *generator* that maps interlingua representations into sentences. In the context of a large multilingual project such as C-STAR or NESPOLE!, this has the attractive consequence that each research group can implement analyzers and generators for its home language only. There is no need for bilingual teams to write translation rules connecting two languages directly. A further advantage of the interlingua approach is that it supports a paraphrase option. A user's utterance is analyzed into the interlingua and can then be generated back into the user's language from the interlingua. This allows the user to confirm that the system produced correct interlingua for their input utterance (that is, whether it has correctly understood the sentence prior to translating it, much as a human translator may do). Figure 5 illustrates interlingua-based MT.

The main principle guiding the design of the interlingua is that it must abstract away from peculiarities of the source languages in order to account for MT divergences and other nonliteral translations (Dorr 1994). In the travel domain, nonliteral translations

Figure 5.   Interlingua-Based Machine Translation between Multiple Languages

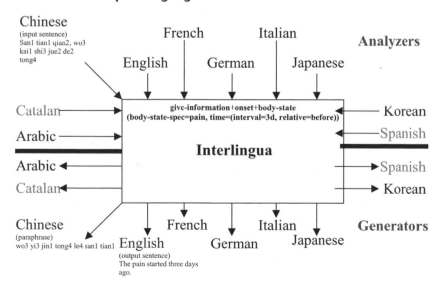

may be required because of many fixed expressions that are used for activities such as requesting information, making payments, and so on. Similarly, in medical assistance, formulaic expressions are often used when eliciting medical information from a patient or suggesting treatments. The interlingua must also be designed to be language-neutral and simple enough so that it can be used reliably by many MT developers. In the case of the interlingua systems described here, simplicity was possible largely because the developers were working within task-oriented limited domains. In a task-oriented domain, most utterances perform a limited number of *Domain Actions* (DAs) such as requesting information about the availability of a hotel or giving information about the price of a flight. These domain actions form the basis of the interlingua, which is known as the *Interchange Format*, or IF.

The IF defines a shallow semantic representation for task-oriented utterances that abstracts away from language-specific syntax and idiosyncrasies while capturing the meaning of the input. Each utterance is divided into semantic segments called *semantic dialog units* (SDUs), and a *Domain Action* (DA) is assigned to each SDU. A DA consists of three representational levels: the *speech act*, the *concepts*, and the *arguments*. In addition, each DA is preceded by a *speaker tag* to indicate the role of the speaker. The speaker tag is sometimes the only difference between the IFs of two different sentences. For example, "Do you take credit cards?" (uttered by the customer) and "Will you be paying with a credit card?" (uttered by a travel agent) are both requests for information about credit cards as a form of payment. In general each DA has a speaker tag and at least one speech act optionally followed by a string of concepts and/or a string of arguments. In example 1 in fig. 6, the speech act is *give-information*, the concepts are *availability* and *room*, and the arguments are *time* and *room-type*. Example 2 shows a DA which consists of a speech act with no concepts attached to it. Finally, example 3 demonstrates a case of DA that contains neither concepts nor arguments.

These DAs do not capture all of the information present in their corresponding utterances. For instance they do not represent definiteness, grammatical relations, plurality, modality, or the presence of embedded clauses. These features are generally part of the formulaic, conventional ways of expressing the DAs in

Figure 6. Examples of Travel Domain Spoken Utterances and Their Interlingua Representations.

Example 1: On the twelfth we have a single and a double available.
   a:give-information + availability + room (room-type = (single & double),time = (md12))
Example 2: And we'll see you on February twelfth.
   a:closing (time = (february, md12))
Example 3: Thank you very much
   c:thank

English. Their syntactic form is not relevant for translation; it only indirectly contributes to the identification of the DA.

Language Analysis and Generation

In interlingua-based translation systems, translation is performed by analyzing the source language input text into the interlingua representation and then generating a string in the target language. Among these, analysis of the source language is the more challenging and difficult task. The richness of language provides humans with a wide range of ways to express the same basic concept. The same idea can be expressed in many different ways. For example, a doctor querying a patient for the location of a pain or injury could express this using a variety of sentences such as: "*show me where it hurts*" (imperative, command), "*where does it hurt?*" (direct question), "*can you show me where the pain is located?*" (indirect question), "*does it hurt here?*" (yes/no question), etc. Achieving very high levels of coverage of such variations is extremely challenging, even in limited domains. Furthermore, the inherent ambiguity of language is a major obstacle to accurate analysis of meaning. As the coverage of an analysis system increases to cover more and more variety of vocabulary and structure, ambiguity becomes more pervasive and the identification of the *correct* meaning becomes significantly more difficult. These problems become even harder yet when dealing with analysis of spontaneous spoken language input. The major additional issues that must be addressed are the disfluent nature of conversational spoken language, the unique grammatical characteristics of spoken language, and the lack of explicit punctuation or even clearly marked sentence boundaries. The imperfect capabilities of speech-recognition systems further exacerbate these problems, since some words in the input may have not been recognized correctly. Analyzers for spoken language must therefore be "robust" in the sense that they must be capable of extracting the main meaning expressed in spoken utterances, even when this meaning is embedded in a noisy and imperfect input utterance.

Target-language generation is more straightforward than analysis. Whereas analysis must handle the variation in language in expressing the same concept, generation can suffice with only a single appropriate text generation for any given meaning representation. Moreover, since we have control over the generated text, it can be designed to be fluent and grammatical. Appropriate punctuation and even prosodic markers can be inserted within the generated text to help produce better pronounced synthesized speech that is more understandable and natural sounding. General text generation frameworks such as GenKit (Tomita and Nyberg 1988), which were originally designed for text-to-text machine translation, have for the most part been equally suitable for target-language generation within speech translation systems and have been used extensively in the various speech-to-speech translation systems developed at CMU.

DEPLOYMENT OF PORTABLE SPEECH-TRANSLATION SYSTEMS

Traditionally, speech-to-speech translation systems have required substantial computing power. Speech recognition benefits from fast processors with ample computer memory. Translation too, both knowledge-based and statistical, requires significant computing power. Concatenative speech synthesis also generally improves with large databases. Hence, best results are obtained when each process is run on a separate fast processor on a local area network.

However the best form factor for use of such systems in front-line medical and refugee situations is a small and portable device. A cell phone connecting to a central service can be a possibility, but cell phone coverage and quality of transmission present considerable additional challenges. An alternative is to miniaturize the processing, compromising some accuracy for speed and memory size, and through clever engineering reduce the memory footprint of the software, as well as take approximate faster methods versus more exact slower ones.

In our work we have developed scalable systems that can run on large servers and also on consumer PDAs (personal digital assistants like the HP Ipaq). In porting speech translation systems to handheld computers (Waibel et al. 2003), we must modify a number of key points in the design. Such handheld computers are much less powerful than standard desktop or laptop computers. Although at first Moore's Law, which says that computers will double in power every eighteen months, may be thought of as a long-term savior, we find the actual limiting factor is usable battery power. Batteries improve much more slowly than computer or memory chips. Hence, our handheld speech-to-speech translation engine has been specifically designed for low-power-consumption chips whose instruction sets exclude floating-point computations. These require significant adjustment to our algorithms, including clever approximation techniques to replace more exact computations.

Additionally the small form factor introduces issues with usability in the field. The system must be light enough to carry and fast enough to be useful. Although better quality audio can be obtained with a head-mounted microphone, this can be impractical in a medical interview or refugee-processing scenario. Instead, we use a built-in microphone and must cope with its poorer quality and ambient noise pickup.

Since the primary user of the system will be a health worker or other humanitarian aid specialist not necessarily versant with computing or translation, we must design the interface and functionality accordingly. We do give very brief training on system activation, rebooting, and usage, such as speaking in short clear sentences, which we found enhances the system's performance (Frederking et al. 2002).

Illustrative Scenario: Famine Relief in Somalia

Consider a hypothetical scenario where a developing humanitarian crisis calls for a U.S.-led international relief effort: For illustration purposes, let us say that in 2009, after a period of prolonged strife, a consensus government emerges in Somalia, capable of maintaining a certain level of stability. However, the combination of destroyed infrastructure and a drought season are threatening widespread famine again, but this time due to the relative stability, a sizable relief effort starts to be planned for deployment in three to four weeks. However, lack of trained English-Somali translators poses a major potential impasse. A quick search identifies a handful of individuals, three of whom are willing to help, but two are rather elderly (Somali expatriates, now retired professionals in the United States), and therefore cannot be safely deployed.

However, all three fluent bilinguals can participate in developing a bidirectional English-Somali speech-to-speech machine-translation system, based on the new PTRANS (Portable TRANSlator) technology just completed in the laboratory after a period of stable funding. The three ex-Somalis are then asked to contribute towards teaching the essence of Somali to PTRANS, both written and spoken. The willing and able individual will later also join the relief deployment as the central translator to help broker agreements with local leaders—though she cannot be in multiple places at once, and therefore more routine translations will be assigned to PTRANS. The first decision is that the translation system will need to focus on the domains of primary medical care (doctor-patient interviews, inoculations, and so on) and in the logistics of food distribution (including roads, directions, warehousing, instructions on delivery). It would not be feasible to create a general purpose speech-to-speech translation system in four weeks.

In order to train a Somali speech recognizer, several hours of transcribed recorded speech are needed from multiple individuals. All three contribute their speech and transcription, and later a few more Somali speakers are located and asked for a few hours to complete the task. Both male and female speakers are required to train the speech recognizer adequately. Speech synthesis only requires the recordings of one clear speaker (two if both male and female voices are desired), and one of the two retired individuals is selected since he speaks a well-accepted Somali dialect clearly.

Training the Machine Translation part of PTRANS requires a bit more involvement from the Somali-English bilinguals. One is tasked to lexical issues, checking the common words in the electronic bilingual dictionary and establishing the correspondences between Somali and English inflections. The other two are asked to translate the linguistically balanced elicitation corpus (see section 3 of this chapter), using the elicitation tool for word and phrase alignments. After two weeks of elicitation, the transfer rule learning method (section 3) extracts and generalizes candidate transfer rules, which are then used to produce test translations of new phrases and sentences. The Somali-English bilinguals check these translations for accuracy, noting which translations are incorrect and classifying the errors. The learning system uses these corrections to repair and augment the transfer rules, producing a working translation system in the domains of primary care and food-distribution logistics.

The last week is used to integrate, test, and further refine the three phases: speech recognition, translation, and speech synthesis. Several dozen handheld PTRANS devices are loaded with the new English-Somali system and distributed to the members of the deployment team. With the departure of the team to Somalia, the PTRANS system work continues for several more weeks, improving the coverage and accuracy of both the speech and translation components. Data is relayed back from field units (uploaded at the end of each day) on sentences and phrases the system was asked to translate, especially ones where it may have failed to do so correctly. This data permits further refinement of PTRANS to actual field conditions during the deployment.

BIBLIOGRAPHY

Black, A., K. Lenzo, and V. Pagel. 1998. "Issues in Building General Letter to Sound Rules." In *Proceedings of 3rd ESCA Workshop on Speech Synthesis,* 77–80. Jenolan Caves, Australia.

Black, A., and K. Lenzo. 2000a. Building synthetic voices: The FestVox project. http://www.festvox.org.

Black, A., and K. Lenzo. 2000b. "Limited Domain Synthesis." In *Proceedings of ICSLP-2000.* Beijing, China.

Brown, P., J. Cocke, S. Della Pietra, V. Della Pietra, F. Jelinek, J. Lafferty, R. Mercer, and P. Roossin. 1990. "A Statistical Approach to Machine Translation." *Computational Linguistics* 16, no. 2, http://www.aclweb.org/anthology/J90-2002.

Dorr, B. 1994. "Machine Translation Divergences: A Formal Description and Proposed Solution." *Computational Linguistics* 20, no. 4: 597–633.

Frederking, R., A. Black, R. Brown, J. Moody, and E. Steinbrecher. 2002. "Field Testing the Tongues Speech-to-speech Machine Translation System." In *Proceedings of LREC*. Las Palmas, Canary Islands.

Hutchins, W.J., and H. L. Somers. 1992. *An Introduction to Machine Translation*. London: Academic Press.

Hutchinson, R., P. N. Bennett, J. G. Carbonell, P. Jansen, and R. Brown. 2003. "Maximal Lattice Overlap in Example-based Machine Translation." Technical Report CMU-CS-03–138/CMU-LTI-03–174 (June).

International Phonetic Association. 1993. IPA: The International Phonetic Association (revised to 1993)—IPA Chart, Journal of the International Phonetic Association 23.

Lavie, A. 1996. GLR* : "A Robust Grammar-focused Parser for Spontaneously Spoken Language." Ph.D. diss., Technical Report CMU-CS-96–126, Carnegie Mellon University, Pittsburgh, Pa.

Lavie, A., C. Langley, A. Waibel, F. Pianesi, G. Lazzari, P. Coletti, L. Taddei, and F. Balducci. 2001. "Architecture and Design Considerations in NESPOLE!: A Speech Translation System for E-commerce Applications." In *Proceedings of HLT-2001 Human Language Technology Conference*. San Diego, Calif.

Lavie, A., S. Vogel, L. Levin, E. Peterson, K. Probst, A. Font Llitjós, R. Reynolds, and J. G. Carbonell. 2003. "Experiments with a Hindi-to-English Transfer-based MT System under a Miserly Data Scenario." *TALIP* (ACM Transactions on Asian Language Information Processing) 2, no. 2 (June): 143–63.

Levin, L., A. Lavie, M. Woszczyna, and A. Waibel. 2000. "The JANUS-III Translation System." *Machine Translation* 15, no. 1–2.

Maskey, S., A. Black, and L. Tomokiyo. 2004. Bootstrapping Phonetic Lexicons for New Languages. In *Proceedings of ICSLP-2004*. Jeju, Korea.

Probst, K., L. Levin, E. Peterson, A. Lavie, and J. G. Carbonell. 2002. "MT for Resource-poor Languages Using Elicitation-based Learning of Syntactic Transfer Rules." *Machine Translation* 17, no. 4: 245–70.

Schultz, T., and A. Waibel. 2001. "Language-Independent and Language-Adaptive Acoustic Modeling for Speech Recognition." *Speech Communication* 35, no. 1–2 (August): 31–51.

Schultz, T. 2002. GlobalPhone: "A Multilingual Speech and Text Database Developed at Karlsruhe University." In *Proceedings of the International Conference on Spoken Language Processing (ICSLP-2002).* Denver, Colorado.

Tomita, M. and E. H. Nyberg. 1988. "Generation Kit and Transformation Kit, Version 3.2: User's Manual." Technical Report CMU-CMT-88-MEMO, Carnegie Mellon University, Pittsburgh, Pa.

Tomokiyo, L., A. Black, and K. Lenzo. 2003. "Arabic in my Hand: Small-footprint Synthesis of Egyptian Arabic." In *Proceedings of Eurospeech 2003.* Geneva, Switzerland.

Vogel, S., F. J. Och, C. Tillmann, S. Niessen, H. Sawaf, and H. Ney. 2000. "Statistical Methods for Machine Translation." In *Verbmobil: Foundations of Speech-to-Speech Translation,* ed. Wolfgang Wahlster, 377–93. Berlin: Springer Verlag, http://www-i6.informatik.rwth-aachen.de/Colleagues/och/VMBUCH.ps.

Waibel, A., A. Badran, A. Black, R. Frederking, D. Gates, A. Lavie, L. Levin, K. Lenzo, L. Mayfield Tomokiyo, J. Reichert, T. Schultz, D. Wallace, M. Woszczyna, and J. Zhang. 2003. "Speechalator: Two-way Speech-to-speech Translation on a Consumer PDA." In *Proceedings of Eurospeech 2003.* Geneva, Switzerland.

Woszczyna, M., N. Coccaro, A. Eisele, A. Lavie, A. McNair, T. Polzin, I. Rogina, C. P. Rosé, T. Sloboda, M. Tomita, J. Tsutsumi, N. Aoki-Waibel, A. Waibel, and W. Ward. 1993. "Recent Advances in JANUS: A Speech Translation System."

Every message needs a messenger. The speed at which the message is delivered, the number of people it reaches, and the effect it has can change lives.

In natural disasters the earliest hours are the most vital. Few die after the opening hours. The knowledge of where to go, what and where to avoid, what to send and where to send it saves lives.

In man-made crises sometimes it is better to stay than to move. If you move, you need to know to where and how and what dangers you will encounter. When you return you need to know what awaits you, mines or booby traps, friends or foe.

In interethnic violence the media has proved to be both a force of evil and of power. Radio stations have spewed forth venom and bias, TV images have fueled insurrection, and the written word has published false gospels. Governments and factions have manipulated the news to their own ends. Even the most independent media stations have struggled to remain impartial in their output.

In the Balkans and in Central Africa, TV and radio stations are attempting to reunite communities by using soap operas that emphasize the common bonds and older values that existed before the violence. The difficulties of return and reintegration, forgiveness and reconciliation are played out in tightly scripted daily episodes.

The obscene aftermath of indiscriminate mine laying, the loss of limbs, the loss of livelihoods, and the loss of self-respect can be mitigated by an awareness of the danger. Mine Awareness Programmes can save more lives than Mine Clearance activities. Posters and banners in schools and community buildings, advertisements in newspapers, warnings on radio and television can prevent children from playing and farmers from planting and tilling in suspect areas.

In crises people need to know more than ever what is happening and who is winning, who is causing the pain and who is curing it. Rumor spreads faster than fact. It is more difficult to correct misinformation and disinformation than it is to disseminate truth. If desperate people believe that transport is waiting to take them away to safety, if starving people believe that food is to be distributed and both of these are untrue, the blame will be placed upon the shoulders of the agency not on the authors of the untruth. The reaction and response may range from distrust to destruction and rioting.

The major media companies are eager to cover "breaking news" crises, their pictures and their comments are accepted as the facts, and when general public interest wanes or the news becomes routine, they move on, leaving an information vacuum in the very place where news coverage is needed most. Switching off the camera lights falsely implies that this crisis is over, whereas for the local population the long journey to recovery has hardly begun.

—Larry Hollingworth

Media for Reaching Large Audiences

Joseph Bravman, Ph.D.

INTRODUCTION

Humanitarian efforts need modern communications both from the perspective of those engaged in the humanitarian work and those who are the recipients. Modern wireless telecommunications offer many new solutions that may be exploited by matching requirements with the characteristics of the planned and available capabilities of systems. These may be divided into a number of categories either from the perspective of the application or from the basic features of the technology.

These wireless technologies, especially those that are space based, can be very powerful, as they tend to require much less ground infrastructure. This is a critical attribute, whether it is needed because of natural disasters, war or terrorism, or an underdeveloped economy.

This chapter will focus on the use of some newer satellite-based capabilities that were created during the latter 1990s and those that are likely to see growth in the early twenty-first century. Specific emphasis will be given to broadcast systems and their ex-

Figure 1. Characteristics of Wireless Technologies

	One way (broadcast)	Two way (dedicated)	Two way (connection based)
Satellite	DTH-TV, SDARS	VSATs, etc.	LEO and GEO phones & data terminals
Terrestrial	Radio & TV stations	Point-to-point microwave (includes 802.11)	Cellular phones (GSM/GPRS, etc.)

panded application to humanitarian purposes. In the figure above this is shown in the first column. Specifically, the use of satellite-based broadcast offers unprecedented ways to reach large numbers of people with a broad spectrum of multimedia information while also delivering information simultaneously to smaller, defined groups.

While these satellite broadcast systems were originally marketed as satellite-based alternatives to familiar radio and TV broadcasts, the systems that were fielded as well as those that will be proposed are intrinsically digital so that other media or applications can easily be adapted—bits are bits. In one case the systems' architects insured that a low-cost data capability was present in parallel with the traditional audio features, thus making flexible multimedia applications much easier to implement. As these systems are fully digital the information can be efficiently distributed as data files, images, geo-coded graphical data, or text. And they can be stored, sorted, manipulated, displayed, and printed based on specific applications programming. Such information can be brought to individuals or kiosks, or made available for redistribution.

A broadcast modality has the ability to present a very low cost to the user due to the large number of receivers built and the ability to divide the cost of content and transmission by the number of recipients. In addition, the lack of a transmitter at the user terminal further reduces cost, consumes less power, and enables a simpler regulatory and operational framework.

Sample broadcast applications range over a broad spectrum that covers:

- High quality weather—real-time and predictive
- Entertainment including low bandwidth video
- Distance learning and training
- Database updates
- Disaster (natural) and threat (man-made) alerts
- Magazines, newspapers, and media
- Health—medical data and HIV/AIDS education
- Fleet operations and tracking displays
- Law enforcement

Satellite Broadcast Systems

Over the past several years some noteworthy developments have occurred that provide low cost, nearly ubiquitous methods to distribute multimedia to large numbers of people on a national or even continental basis throughout most of the world. The most successful and visible of these is satellite TV broadcasting, which today peppers rooftops with half-meter Ku-band (12 GHz) dishes. These installations are also commonplace in less affluent or more rural settings, and often can provide Internet connectivity. Deployment has slowed over the past five years due to the general downturn in dot.com and high-tech investments coupled with the business failure of many LEO (Low Earth Orbit) satcom projects. However, the geostationary satellite projects have generally fared better, due in part to their regional focus and the need for as few as one satellite to create a service.

The wireless (radio frequency) spectrum is controlled by the ITU (International Telecommunications Union), headquartered in Geneva, Switzerland, and its member state regulatory bodies, who allocate frequencies over three defined regions of the globe. These allocations suffer from many of the historical and political issues that land zoning boards face. The allocation decisions are also based on physical laws, and like the zoning analogy, these decisions are likely to remain along with legacy systems that operate with their framework.

In general the lower portions of the microwave spectrum (1–4 GHz) are employed more for mobile communications, where these signals are more weather immune and may be received with small, nearly omnidirectional antennas. The bandwidth available, however, is lower than at higher frequencies (6–30 GHz) that are more suitable for wideband and fixed-site applications. They require larger aperture antennas (one-half to two meters) and accurate pointing, and they often suffer outages during heavy precipitation unless precautions such as geographic diversity are taken.

Satellite TV has become a well-established commercial success in many parts of the world; however a less well-known development is also taking place with a technology called Satellite Digital Audio Radio or SDARS. Unlike their TV cousins, these systems were designed at the onset for mobile users rather than primarily fixed installations. In the United States the target market was automobiles, and those characteristics drove the requirements. In a vehicle power was not an issue, but the rapid motion of the car

Figure 2.   Wireless Frequency Utilization

	1–4 GHz (L & S-Band)	4–8 GHz (C-Band)	12–30GHz (Ku & Ka-Band)
Satellite	Satellite phones, SDARS	VSATs, trunking	VSATs, DTH-TV, Internet
Terrestrial	Cellular phones, 802.11		Line of sight Microwave

often placed obstacles between the satellite and the antenna (as severe as underpasses). To relieve this loss of service quality the systems employed more than one satellite for spatial diversity and, when even that proved inadequate, added hundreds of ground towers across the United States primarily in urban areas. XM and Sirius are the two licensed U.S. operators.

Specific frequencies with the L- and S-band portions of the radio spectrum (corresponding to 1.5 or 2.3 GHz) were chosen because they fit the requirements described above. Typically, an antenna for these services can be nearly omnidirectional and is about the size and shape of a hockey puck. The satellites transmit a relatively wideband signal (1.5 to 4 Mb/s) and subdivide it using time-division multiplexing to produce about one hundred channels that carry digitized audio or, as will be discussed later, digital data. These systems were deployed for regional use that favored geostationary orbits (although Sirius Satellite Radio uses high earth orbit inclined elliptical orbits which are synchronized to the earth's rotation).

Internationally, the system that was deployed by WorldSpace flowed from a different paradigm. The goal was to support highly portable and very inexpensive radios. This goal was set to bring a broad range of audio programming to underdeveloped regions of the world that often lacked disposable income or a reliable source of electrical power. Thus, simplicity and low power were dominant requirements. Single geostationary satellites cover each region, and no towers are deployed. The receivers must be able to be run from small conventional batteries (AA or C cells). The figure below depicts the coverage area provided by a pair of geostationary satellites, one serving primarily Africa and Europe and the other Asia and a portion of the Pacific to western Australia.

Figure 3.   Satellite Coverage Map

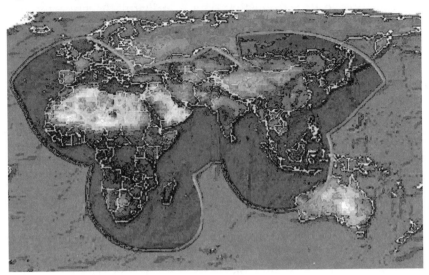

High volume production rapidly lowered the cost to the point where even the unsubsidized price is in the one-hundred dollar range and multiple suppliers exist from many of the global consumer electronics manufacturers.

The broadcast concept transforms the satellite operators and their strategic partners from owners of space-based infrastructure, such as is more prevalent with the C-band and Ku-band transponder leases and VSATs, into media providers. The result is that if the content can be produced, it can be distributed in multicast or broadcast fashion. Furthermore, the business models range from pure subscription service to a free service subsidized by advertising, philanthropic, religious, or governmental funding.

Initially, these systems were viewed by their owners as audio broadcast platforms (as the acronym SDARS would suggest); however, as the technology is fundamentally digital it became apparent to these companies and their strategic partners that other digital media could be transported with appropriate hardware and/or software developments. Applications have been demonstrated by a number of companies that range from highest quality graphical weather to compressed video.

The broadcast paradigm lowers the user cost for content by

dividing its total cost by the number of users who occupy the same user group. To the extent that data is individually requested, the cost would be borne by the requestor. This permits a straightforward economic model to be developed. While the primary modality is broadcast, such systems can be coupled with some type of asymmetrical path (perhaps non-real-time) that can be used to request or customize the transmissions. Broadcast also allows information to be received simultaneously by the entire authenticated group, which is ideal for emergency notification or information sharing.

The key point here is that a capability now exists to distribute audio (in native languages), highly compressed video, text, databases, software, or graphical data to large and dispersed user groups at extremely low cost. Sophisticated computing power exists in modern laptops, tablet PCs, and pocket PCs/PDAs, which allows software to handle authentication, decryption, decompression, and geographic filtering that increases user friendliness and relevance. In cases where the user is untrained, it also can automate the reception, formatting, and display.

Figure 4 shows in block diagram form the way in which the information is assembled in a data center and transferred to the satellite uplink facility. The satellite then broadcasts the data to all authenticated users within its footprint over a designated data channel. The simultaneous reception of GPS data can add an op-

Figure 4.   SDARS Block Diagram.

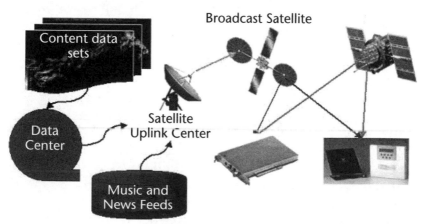

tional feature allowing the user device to derive location information to plot his relative position or to preselect geographically relevant data.

SDARS ENABLED APPLICATIONS

The following is a partial list of application areas and is provided only to stimulate the reader. In general, if valuable content can be created, it can be broadly transmitted and turned into user-friendly information with a variety of PC-based devices and application-specific software. Several of the early applications that are discussed have paved the way for other applications by incurring the costs or underwriting the technological development that is required. Each application area contains companies or experts that have or are developing IT-based solutions that would benefit from the addition of a data feed that carries their information content. In a few cases adaptations are required to convert the need for a duplex (two-way) link to broadcast or at least a highly asymmetrical connection.

High Quality Weather: Real-Time and Predictive

The weather affects everyone's life in many ways. Weather may be sent via audio or as data sets. In the first case it is available to a broad set of people whose only requirement is language compatibility, although a listener may have to wade through a long period of irrelevant information. Also the information must be of low resolution to permit only a verbal description. Nonetheless, audio broadcasts have their place, especially with less well-equipped people.

Currently such broadcasts are available over WorldSpace in English and provide forecasts covering seventy-seven zones from latitude 54 north southward, including Europe, the Mediterranean, and North and West Africa. Shipping and Inshore Waters forecasts from sources such as the UK Met Office are converted to speech and then transmitted every thirty minutes. Other weather topics are also included for educational purposes. Long-term forecasts look ahead from twenty-four hours to five days.

On the digital side, however, a broader and more powerful range of high quality, high resolution data sets may be broadcast that may be interpreted or formatted by the display and PC-based processor. If the data is sent to a PC-based device, it can be unpacked into files, and these can be sorted based upon the application, location, or other criteria. Once sorted, the software can present the information in a useful and familiar manner to that class of users. The same precipitation map data sets may be processed and displayed differently for use by a fisherman, an airplane pilot, a farmer, or for flood/storm disaster management. Spatial resolution can be as fine as 1 km, and update rates for new information are as frequent as every five minutes. Some examples of graphical weather are shown in Fig. 5.

Figure 5.   Graphical Weather Examples

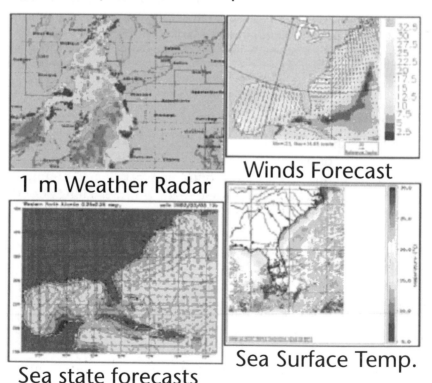

1 m Weather Radar

Winds Forecast

Sea state forecasts

Sea Surface Temp.

Entertainment Including Low Bandwidth Video

There are a number of other dimensions to entertainment. First, content can be oriented to provide ethnic diversity to local programming. Second, entertainment can help solve the common problem of reaching an audience with a controversial message or an important, but boring, one. Such a message can be inserted into a more popular or acceptable genre. For example, a rock and roll station can deliver messages on drugs or birth control issues. The media also has the potential to enlighten populations about alternative perspectives or a more balanced view of other nations.

Software compression technologies have made it possible to show respectable quality video clips over the more limited bandwidth of SDARS systems, especially if they are stored for non-real-time viewing. The use of smaller displays also works in favor of limiting the required bandwidth to as little as 128 kbps.

Distance Learning and Training

Traditional distance education platforms rely on books, tapes, and other materials that may be expensive to produce and difficult to deliver. Internet-dependent approaches require Internet access for delivery that may be slow or not even available. Video conferencing involves very costly equipment and communications channels.

Satellites can broadcast such materials very inexpensively and simultaneously to many sites. An added dimension combines data and audio transmissions. Instructor-led live lectures and accompanying PowerPoint presentations can be broadcast directly to students' PCs at a scheduled class time over a satellite channel. Students hear the live audio of superior teachers, follow along with the presentations, and experience real-time data updates as the teacher works through the class.

A less elaborate approach would allow downloads of significant amounts of presentation materials (text, graphics, PowerPoint, and so on) for archive; display, or printing. Whole books could be transferred in this way using the same bandwidth that would be required for speech. Presentations, lesson plans, and other multimedia materials are easily delivered to students, complementing and expanding the classroom lectures.

If the student site has Internet access or an asymmetrical back channel, it is possible for students to ask questions during the class in a chat or VOIP (Voice Over Internet Protocol) mode.

Database Updates

Data-casting is when there is a need to transmit large volumes of information to multiple locations, especially in areas where traditional communications infrastructures are expensive to use, unreliable, or nonexistent. Such data sets can be data bases and lists, transaction updates, logistics, and software updates, among others. The recipients are generally classified as a closed-user group. This group could be widely dispersed within the beam of the satellite, and each user would simultaneously receive their data.

Once created, this closed-user group would source sets of files that would be sent on a scheduled or queued basis, often daily. The data would be encrypted and authenticated only for group members. Relatively high data rates (up to 128 kbps) would be used in a burst mode to minimize the time required to download a file. A one megabyte file would take about one minute to transfer. Since the broadcast channel is unidirectional, there is no way to acknowledge that a recipient has received an uncorrupted set of data. Thus, the data must be sent redundantly or more than once if it is necessary that it be received without corruption.

Since an individual group would only use the channel for a small percentage of time, data from diverse users would be aggregated onto a single channel or perhaps interspersed with other content.

Disaster (Natural) and Threat (Man-Made) alerts

When disaster strikes there are many messages that need to be relayed to those affected or those who are empowered to provide assistance. Broadcasting to inexpensive receivers offers an attractive means to deliver such information. Data could show the geographic areas that are dangerous or where help is being dispensed. Inventories of food, supplies, or medicines could be sent and updated. Immediate warnings including detailed evacuation instructions could be distributed.

Given the priority nature of these crises, the data would be pushed to the top of the queue. Also the data could exist in several forms based on need to know and the user's ability to display it. Often local information is inaccurate, which leads to chaos or violence. The ability to broadly and rapidly distribute accurate and relevant information to the scene is a key to improving the situation.

Magazines, Newspapers, and Print Media

In a similar manner to the method of disbursing books and distance learning materials, other media can be distributed. Newspapers, newsletters, magazines, and even books can be broadcast and either displayed or printed. Prominently placed kiosks can further amplify the extent to which large numbers of people have access to this media. For people who are far from home, the ability to get familiar newspapers or magazines while in remote areas is a big morale builder.

Health: Medical Data and HIV/AIDS Education

Several nations currently have programs to sponsor audio broadcasts that educate the population to halt the spread of AIDS/HIV. Public health applications can also include information about locations for vaccination programs, as well as updates to medical inventories, which was discussed in the more general case of logistics data.

New medical procedures can be disseminated to remote areas in ways similar to distance learning.

Fleet Operations and Tracking Displays

Situational awareness information can be sent to provide graphical locations of vehicles or other tracked objects. This application also has characteristics that overlap logistics and other areas, including law enforcement.

Law Enforcement

Applications range from locating friendly support units or supplies to sending mug shots of criminals or reports of suspected
criminal activity. Police, border patrol, and other security forces
can get alerts and up-to-date threat information. While such
groups have various types of radio systems, these usually have limited bandwidth and may not have countrywide information feeds.
Again many applications have similar characteristics to those discussed above. Such information can be displayed as pictures,
maps, or text.

RECEIVER AND DISPLAY EQUIPMENT

A number of configurations are available commercially that can
be used to receive data transmissions. These are illustrated in fig.
6, which shows the use of receivers in common user configurations. Depending on the application, the equipment may be packaged for fixed or highly portable personal use.

Figure 7 shows a range of receivers that typically cost about one
hundred dollars each. One is a PCI card and can be inserted into
a standard desktop PC. A fifty-foot antenna cable is provided that
can be further extended with use of a small in-line amplifier.
Other receivers are stand-alone and interface using USB cables.
One of the smallest receivers is slightly larger than a deck of cards
and is so simple that it is even powered through the USB port
with only an antenna port as the other connection. All other functions are managed by the PC. Many basic software drivers are
available and or have been customized to support specific applications by data service providers. As usage proliferates, other optimized hardware and software will emerge at affordable prices.

As the automotive market for satellite broadcast systems becomes successful in the United States through the efforts of XM
and Sirius, other regions (initially Europe) will likely follow suit.
This will give rise to a more robust portable system, bringing with
it a set of developments that should benefit humanitarian adaptation of these products and services.

Figure 6. Typical Configurations

Figure 7. Typical Receiver Equipment

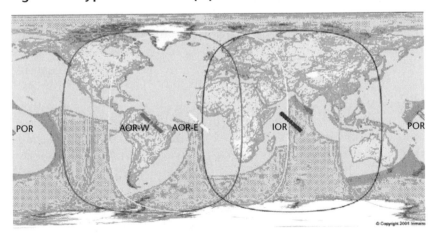

FUTURE DEVELOPMENTS

These discussions reflect current or very near term capabilities. A study of the satellite communications industry has shown it to be a field that has never seen steady progress, but rather tough times interspersed with leaps of progress and success. There have been periods when access to space, capital, or technical hurdles created delays. The promise of space-based ubiquitous high-speed Internet has eluded us from the early promise of Teledesic in the 1990s. However, on average there exists today a broad range of capabilities. Today's satellites are more powerful and able to support less expensive receivers and smaller antennas, although they still require significant capitalization to launch and operate.

Over the longer term further improvements in IT and communications will continue, stimulated by large markets and needs. And while there will be periodic lapses in capital spending for space-based telecommunications, the overall prospects continue to be strong, especially where there have been proven business successes. By way of example, continued improvements in compression techniques, better antennas (including low-cost phased arrays), and spacecraft power and costs will provide new capabilities. L- and S-bands will continue to be favored for mobile applications, with new offerings on the horizon from Inmarsat for laptop-sized two-way data terminals. Ku and ultimately Ka-bands (20 and 30 GHz) will offer enhanced wide bandwidth capabilities. These basic building blocks will become linked into hybrid networks that also employ various terrestrial wireless techniques (GSM/GPRS, Wi-Fi, and others) to provide enhanced local reach at low cost.

These space-based capabilities have the advantage of creating "instant infrastructure" in regions that may lag behind more developed parts of the world or where man or nature has cause massive disruption to the telecommunications fabric. This makes them very powerful tools for reaching the goals of humanitarian efforts and very flexible for technologists to adapt. When properly adapted, technology offers important opportunities for a broad range of humanitarian purposes that can only be optimized by a partnership between the technologist and humanitarian, each of whom know their respective needs and capacities.

GLOSSARY

L-Band	Radio frequencies from 1–2 giga-hertz (GHz)
S-Band	Radio frequencies from 2–4 giga-hertz (GHz)
C-Band	Radio frequencies from 4–8 giga-hertz (GHz)
Ku-Band	Radio frequencies from 12–18 giga-hertz (GHz)
GEO or GSO	Geo-synchronous satellite orbit. When the orbital period is twenty-four hours and the orbit is equatorial, the satellite moves in synchronism to the daily rotation of the earth and as such appears stationary above the equator at a specified longitude. Inmarsat, XM, Thuraya, Intelsat, Panamsat are examples of GEOs.
LEO/NGSO	Low Earth Orbit/Non Geo-synchronous Orbit. LEOs are generally below 850 km above the earth's surface and in highly inclined or nearly polar orbits. As a result it takes a large number of satellites to provide continuous coverage of a region. They often require smaller antennas, as the path distance between the user and the satellite is smaller, and suffer less signal delay than GEOs. Iridium, Globalstar, and OR-BCOMM are LEO systems. Sirius is a high orbit NGSO.
Broadcast	One-way communications from a source to a large number of recipients in a broad coverage region.

SDARS

Satellite-based Digital Audio Radio Systems that use L- or S-bands to distribute digital audio (and potentially data) broadcast services. WorldSpace, XM, and Sirius are SDARS carriers.

Duplex or Simplex

Communications provided in both inbound and outbound directions. Duplex allows this simultaneously as opposed to Simplex, which supports either or alternates between the two path directions.

GSM/GPRS

Global Mobile Standard for cellular telephony. GPRS describes the corresponding data capability for messaging.

WiFi/802.11 a/b/g

Standard for wireless short-range computer communications. The primary standard operates at S-Band (2.5 GHz), although C-Band offers future expansion. The "g" version of the standard expands the 11 M bits per second data rate to 54 Mb/s.

USB

Universal Serial Bus that connects common peripherals to PCs.

Omnidirectional

An antenna type that is able to receive signals from any (omni) direction and therefore does not require pointing; however, this flexibility is achieved at the expense of sensitivity.

VSAT

Very Small Aperture Terminal. A satellite terminal that uses a small dish antenna, usually at Ku-Band and about $1/2$ meter in diameter, to provide two-way communications, often as part of a private network.

Real-time	Instant and continuously available communications as opposed to store-and-forward communications or those that are not continuously available for use.
DTH	Direct to Home. Often used to describe satellite TV broadcasting.
Dedicated	A permanent telecommunications connection such as a leased line.
Connection-based	A telecommunications method whereby a virtual connection is established by the caller for the duration of the information transfer, like placing a telephone call.
Line of sight or Point-to-point	(LOS) A direct unobstructed path between the antennas is required.
Data rates	The rate at which data is passed over the circuit in bits or bytes per second (a byte is equal to 8 bits). Satellite voice circuits generally provide 2.4K to 9.6K bits/second in data mode. Satellite data circuits can support rates up to about 128 Kbps with growth anticipated to higher rates in the future. Broadcast systems transmit millions of bits per second and subdivide this rate into effective channels.
TDM	Time division multiplexing. A system by which a high data rate is subdivided into channels that are sequentially sent in accordance with a predetermined schedule that is continuously and rapidly repeated.

SUPPORT TECHNOLOGIES

In a single week the killing of five MSF (Médicins Sans Frontières) staff in Afghanistan, gunned down in their vehicle as they left their office at the end of the day to travel home, has shocked and saddened the aid community. At the same time in another troubled area, fifteen UN employees were taken hostage in Dhafur in Sudan. Over the last ten years more than 240 UN civilian employees have died as the result of violence. Six members of ICRC in Chechnya were murdered in one incident in 1996. In July 2003 eleven local aid workers were taken hostage and then killed in the Democratic Republic of the Congo. In June, Annalena Tonelli received the prestigious Nansen Refugee Award from UNHCR for thirty-three years of service to refugees in Somalia. In October she was murdered on the grounds of her hospital. Last year also witnessed in Baghdad the deliberate targeted bombing of the headquarters of both ICRC and the UN.

The 1994 Convention on the Safety of United Nations and Associated Personnel came into force in 1999. The Convention has been ratified by sixty-nine countries, but as the UNHCR's excellent magazine *Refugee* says, with a hint of cynicism and incredulity, "crucially it does not cover UN humanitarian workers in most situations." The following story proves the point.

> I heard the thud of bullets passing through the body of the vehicle. Suddenly I felt warm blood streaming down my neck from my head. . . . The armed men eventually rounded us up and forced us to sit down against the wall of a house. There were around 30 men armed to the teeth. Some were very young . . . one had taken Saskia's spectacles and she could not see any more. I talked to him and he gave them back . . . the last words I heard from her were "They have taken my shoes." The armed men started to withdraw. Another explosion . . . I jumped up and ran. I looked back and saw Saskia on her left side, blood soaking out of her blond hair. Luis was lying dead beside her.
>
> A moment later another UN colleague, Guy, came around the corner . . . with two other colleagues. He shouted "run." We ran . . . all the time expecting to be hit in the back. Leo was hit in her breast and couldn't run any longer. I held her and we followed Guy and Kathleen . . . I had lost quite a bit of blood. I had left Leo under a tree . . . Gunfire erupted again . . . We crossed a river . . . Guy went back with the army to recover the bodies of Saskia, Luis, the Burundi assistant. They also brought back Leo, who had been found by farmers, her wound was dressed and glass and bullet fragments were plucked from my skull.

We flew home with the bodies of our friends in body bags in the aisle of the aircraft. ("Journey into Darkness," Burundi, October 12, 1999, by Christoph Hamm, Repatriation Officer, UNHCR. From *Refugees,* UNHCR magazine, vol. 4, no. 121, 2000).

The bombing of the UN headquarters in Iraq resulted in three separate inquiries. The question fundamental to each was whether or not it is possible to reduce the risk of danger to an acceptable level and if so, how?

The inquiries established that ensuring personal security awareness was the first requirement: the need for all personnel to understand the gravity of the day-to-day threat. The second requirement was the provision of adequate personal security equipment—bulletproof jackets and helmets. The equipment is expensive, but its provision is only one part of the equation. It must be worn to be successful, and this relies on personal discipline.

Traveling to and from workplace and home is a major at-risk function of daily life. The route may be varied, but the start and end point are invariable. Timings and routes can be made flexible, but areas of heavy traffic, roundabouts, underpasses, and bridges all expose the travelers to the threat of a sniper's bullet, an ambush, or an improvised explosive device.

Office and accommodation locations are at greatest risk from vehicle-borne bombing, whether static or mobile, parked and left or driven by suicide bombers. The only safety factor is distance. Can the vehicle be kept away from the building?

But vehicle bombs are not the only threat; terrorists frequently have easy access to rocket-propelled grenades and mortars. These threats have imposed the need for fortress accommodations and fortress offices.

But should aid agencies operate within fortresses? How can the daily life of agencies function behind the fortress? Who will visit these fortresses? How do local neighbors, shopkeepers, and civilian residents whose streets are barricaded and who cannot reach their own homes and shops view these fortresses? In Iraq many agencies moved their offices outside the country to neighboring states. How is this viewed by the beneficiary?

Will the provision of adequate security prevent the face-to-face meetings previously so essential to the conduct of humanitarian assistance? Has the era of the aid worker who is loved, respected, and granted humanitarian space ended? Is the delivery of humanitarian aid now seen solely as a high-risk profession? Have T-shirts and sandals been replaced by flak jackets and helmets?

—Larry Hollingworth

Enhancing Security for Humanitarian Operations through Technology and Information

Joseph V. Braddock, Ph.D.

INTRODUCTION

Much has been written concerning security in humanitarian operations. In that context, security is a concern for both the victims of conflicts and disasters and the humanitarian workers attempting to help them. An excellent summary of these is contained in a chapter by Gerald R. Martone, "Protection Strategies in Humanitarian Interventions" contained in *Emergency Relief Operations* (2003), edited by Kevin M. Cahill, M.D.

In his chapter, Martone explains the possible sources of protection. In simplified form, they are:

Case A)
> Protection accruing to the presence of humanitarian assistance workers and their organizations. Such protection is attitudinal and can extend to both victims and workers.

Case B)
> Protection resulting from valued humanitarian assistance beyond just presence. One might think of this as a quid pro quo circumstance benefiting the workers at some risk or burden for the victims.

Case C)
> Self-protection (as described by Martone), which involves harnessing not only the victims but the effects of "mobilization of shame" from the media and other exposure of the circumstances. When applicable, it helps both victims and workers.

Case D)
> Direct protection, which includes everything from protection by a host government (or a locally ruling element) to legal and educa-

tional means. Formal military or local police protection is part of this approach.

This chapter suggests additional new security dimensions. It describes technology, information, and training that could improve safety and protection for humanitarian workers. Since acquisition, ownership costs, and training require resources, the intellectual approach to addressing these significant collateral matters is to posit the idea of humanitarian organizations and workers as "smart buyers."

This concept applies in several ways. The most straightforward is in effective direct acquisition of technology by the humanitarian organization for its workers. The smart buyer concept also applies to two additional circumstances. It permits the humanitarian organizations to assess the effectiveness and burden of protective means available from third parties and make informed judgments concerning use. Similarly, the same knowledge allows the humanitarian organizations to assess the level of expected protection of indigenous equipment when it is offered.

DISCUSSION OF THREATS, LIMITATIONS OF PROTECTION AND PREPARATION, AND OTHER IMPORTANT CAVEATS

The smart buyer approach is applied to technology, training, information enhancements, and other preparatory activities in all cases because of possible conflict escalation. It applies even when legitimate governments are providing security, protection, transportation, communications, and a generally peaceful environment in which humanitarian assistance is to be delivered. This also includes friendly and helpful response forces that can reinforce local forces.

Realistically, in the worst threat cases, avoidance must be an acceptable strategy. Again, the assumption is made that assistance will be provided even though there is risk. Having thus decided to go forward, what are the threats and risks?

Combatants have a large and diversified weapons inventory available to them. The most prevalent direct-fire weapons are small arms, assault rifles, hand-thrown and rocket-propelled grenades, machine guns ranging from light to heavy, mines, and high explosives used as personal and car/truck bombs. Added to these are indirect-fire weapons such as traditional artillery and mortars. In some instances combat platforms (tanks, infantry fighting vehicles, technical vehicles, helicopters, and older fixed-wing aircraft) can provide more lethal direct and longer-range fires.

In suggesting technologies to improve the safety and security of humanitarian workers, emphasis is given to protecting against the initial categories of direct and indirect fire weapons and even others to a limited extent. Humanitarian workers are not combat soldiers. Personal and some collateral protection and related supporting means are described in the following sections. What must be kept in mind is that escalation in platforms, weapons, and munitions is a possibility that cannot be dismissed. The smart buyer approach is therefore constrained to protection, preparation, planning, training, and operating against the highly proliferated but lighter portion of the threat (which is still highly lethal to humans). The smart buyer concept produces limited but useful improvements. This is also consistent with the general circumstances of "traveling light," which almost universally describes humanitarian operations and workers.

The technologies to be considered in the following sections are:

A) Light armored vehicles
B) Body armor
C) Situation awareness technology
D) Barriers
E) Protective blankets
F) Fire retardant technologies
G) Unmanned aerial vehicles
H) Security assessment software
I) War-gaming and simulation for training.

In each of the following sections, technology in some form is described, and representative sources are attributed along with contact information.

A. Light Armored Vehicles

In general, vehicles offer little or no protection against even the lightest of direct and indirect fire threats summarized earlier. Military combat vehicles are passively and in some cases actively protected. In large units such as divisions, self-propelled vehicles number about five thousand. Even of these, only one thousand are armored in some way. The remainder are protected to the extent of their passive structures. Only over very limited ranges of incoming trajectories is there substantial protection. Bullet fragments, slugs, or shaped charge jets readily penetrate doors, windows, and frames. Only the engine provides significant protection, but the angular protection is a small percentage of the total solid angle subtended.

Three examples of lightly armored vehicles are drawn from a database of protective technologies. In this case, the database is derived from a Force Protection Equipment Demonstration (FPED-IV) held in May 2003 at the Quantico Marine Corps base in Virginia and sponsored by the Office of the Secretary of Defense's Program Manager for Physical Security Equipment. The equipment CD assembled for this event is a major reference for this chapter.

The best case armored vehicle is a military example, the U.S. Army's M1097 A2 vehicle fitted with a protection kit provided by the O'Gara Hess and Eisenhardt Armoring Company. The capabilities achieved are shown in fig. A-1. Two commercial vehicles are also included. They are the John Deere Military Gator and the Chevrolet Suburban Armored (Fig. A-3). These provide differing kinds of protection. The Gator has exposed crew positions, but the vehicle itself is somewhat hard. The Suburban offers protection to crew, passengers, and cargo.

These first three examples highlight the kind of information that is available and the sources for that information. Those experts who support humanitarian work could actually examine the equipment by going to the force protection demonstrations at the Quantico Marine Corps base. These events are scheduled every two years, with the next taking place in 2005. (The PM-Physical Security Equipment office that organizes these events is located at Ft. Belvoir, Va., and can be reached at (703) 704-2416.)

B. Body Armor

As its name indicates, body armor is worn as passive protection. It is effective against small arms, automatic rifles, the lightest of

Figure A-1.

M1097A2 Crew Protection Kit

Category:
Armored and Utility Vehicles
M1097A2 Crew Protection Kit (CPK)
Point of Contact:
Name: Mr. John Mayles
Phone: (513) 881-9899
Fax: (513) 874-2558
Toll-Free: (800) 697-0307

Email: jmayles@ogara-hess.com
Website: http://www.ogara-hess.com

Company Information:
O'Gara-Hess & Eisenhardt Armoring Company
9113 Le Saint Drive
Fairfield, OH 45014
USA

Description:
The O'Gara-Hess & Eisenhardt Crew Protection Kit designed for the M1097A2 chassis, provides a low cost, versatile, interchangeable armor system that can readily be installed on any M1097A2 with minimum modification performed to the host vehicle.

Armoring System includes: M80 Ball Perimeter Armor (doors, rockers, pillars, cowls, windscreen capping and rear panels); M80 Ball High Light Transmission Transparent Armor; 155 mm Fragmentation Protection (Roof) and Contact-detonated Anti-Tank Mine Blast Protection (front and rear).
Cost Data: Contact us for a price quote.

Figure A-2.

John Deere Military Gator

Category:
Armored and Utility Vehicles

Equipment Name:
John Deere 6x4 Diesel Military Gator -
Point of Contact:
Name: Mr. Dan Smith
Phone: (910) 369-4575
Fax: (910) 369-2620
Toll-Free: (800) 358-5010
Cell: (910) 977-1161

Email: smithd@carolina.net

Company Information:
John Deere Company
2000 John Deere Run
Cary, NC 27513
USA

Description:
The John Deere Military Gator, M Gator, is a commercial-off-the-shelf, small tactical/utility vehicle. Based on Army and Marine Corps requirements, the diesel (JP8) powered M Gator is safe, economical, highly mobile, air drop and sling certified and easily transported in fixed and rotary wing cargo aircraft including the V22 Osprey. The M Gator has been extensively tested by the Army, Navy, Marine Corps and Special Operations Command and has a proven track record as a force multiplier. Its low cost and optimum payload to weight ratio (1400/1450 lbs.) provide an immediate and affordable advantage to military and quasi-military operations where the size and cost of the HMMWV can result in a strategic compromise. Over one thousand M Gators are currently fielded at XVIII Airborne Corps, 82nd Airborne Division, 101st Division, the 10th Mountain Division and various Special Operations Units. In addition to the Army uses the National Guard have adopted the M Gator with modifications for their Civil Support Team use.

Cost Data: John Deere M Gator: approximately $14,500.00. Available on GSA Contract.

Figure A-3.

Chevrolet Suburban Armored

Category:
Armored and Utility Vehicles

Equipment Name:
Chevrolet Suburban Armored
Point of Contact:
Name: Mr. Martin Cardenal
Phone: (305) 477-1109
Fax: (305) 477-1139
Toll-Free: (800) 998-2264

Email: gladysg@sq1armor.com

Company Information:
Square One Armoring Services
1435 NW 82nd Avenue
Miami, FL 33126
USA

Description:
SQ1 FAV+ Chevrolet Suburban. Our FAV+ armor package is designed to protect against 7.62 x 51 mm NATO .308 Winchester, 5.56 x 45 mm US M193. The vehicle has full perimeter, roof and floor protection. Various options are available such as ram bumpers, PA systems, and run flat tire inserts.

Cost Data:
For customized and individual quotes, please call Martin Cardenal or Gladys Guillen at 305-477-1109

machine guns, and the smaller or slower fragments of grenades, mortars, and artillery. Bullets, slugs, and fragments designed specifically to pierce armor are still lethal threats. Body armors are sometimes sufficiently hard to trigger some shaped charge variants of dual mode munitions.

Three examples of body armor and armor suits are shown in Figures B-1, 2, and 3. The first two figures display traditional proliferated body armor used by both the military and police. The second example is a lightweight and more expensive version of the example shown in Fig. B-1; it is typically used by fliers.

The third example is the countermine suit and is included for a specific purpose. Mines are a clear and present danger in many humanitarian operations. The expectation is that some other organization will take the responsibility for clearing mines, and that is the desired solution. On the other hand, a mine may be uncovered or an area that is mined may be discovered. In that instance, humanitarian workers may have to act. Having a countermine suit available could be useful, but it is no substitute for training and other expertise to deal with such mines.

Figure B-1.

PPI Line of Body Armor

Category:
Blast/Ballistics/Protection/Mitigation

Equipment Name:
PPI Line of Body Armor
Point of Contact:
Name: Mr. Mark Smith
Toll-Free: (800) 509-9111
Fax: (954) 846-0555
Cell: (703) 980-8465

Email: marks@body-armor.com
Website: http://www.body-armor.com

Company Information:
Protective Products International
1157 Sawgrass
Corporate Parkway
Sunrise, FL 33323
USA

Description:
Spec Ops

This is a vest with one thing in mind...special operations. This design has been used for the last two decades in military and federal operations. It offers front and back plate pockets either top-loading or bottom-loading, over-the-shoulder ballistic protection, adjustable shoulders, overlapping side protection, protected drag strap, inner and outer belly-band system and flap to add groin protection. Front and back ID available if needed. Comes with LBV keepers on the shoulders.

Cost Data: Contact us for a price quote.

Figure B-2.

Air Pro Style Body Armor

Category:
Individual Protective Equipment

Equipment Name:
Air Pro Style Body Armor
Point of Contact:
Name: Mr. Thomas G. Faust
Phone: (610) 375-8549
Fax: (610) 375-4488

Email: bulletpr@bellatlantic.net

Company Information:
T. G. Faust Inc.
544 Minor Street
Reading, PA 19602-2722
USA

Description:
Used by aircrew personnel worldwide. The vest can be used with the Air Ace Survival Vest and the SRU-21/P Survival Vest. FEATURES: One size fits all (four point adjustable quick release straps). Removable Aramid Ballistic fabric panels sewn in waterproof rip-stop nylon. Outer carriers constructed of fire retardant water repellant high strength polyester fabric or cordura nylon. Front and back armor plate pockets and trauma panel pockets. Unique attachable side protection wings for additional protection.

Cost Data:
Contact us for a price quote.

Figure B-3.

The Countermine Suit

Category:
Individual Protective Equipment

Equipment Name:
Countermine
Point of Contact:
Name: Mr. David Pasqualone
Phone: (423) 562-1115 Ext: 112
Fax: (423) 562-1581
Toll-Free: (800) 722-7667 Ext: 112

Email: dpq@pacabodyarmor.com
Website: http://www.pacabodyarmor.com

Company Information:
PACA Body Armor
179 Mine Lane
Jacksboro, TN 37757
USA

Description:
The Countermine Suit enables soldiers to conduct effective countermine operations in various climactic weather conditions while providing enhanced survivability against antipersonnel mine blasts.

PACA's Countermine Suit withstands the effects of water, salt spray, insect repellent, fungi, and mildew without compromising its protective properties.

Cost Data:
Contact us for a price quote.

C. Situation Awareness Technologies

It goes without saying that safety and security are enhanced with information that provides warning and assessment of possible threats. In military jargon, this is called situation awareness. When coupled with appropriate training and preplanning, potential threats may be countered and/or avoided. There are a number of distinct families of technologies that can improve situation awareness. The examples shown in this section are vision devices, since their outputs are readily interpretable with very little training.

Humanitarian operations often take place in poorly developed countries. Threats are enhanced when darkness falls or bad weather occurs. Even in those circumstances where either government forces or a local group are providing security, an independent assessment of external activity is useful.

The examples included here involve two forms of sensors. The first form of sensor is one that images thermal energy. All objects, living and inanimate, radiate thermal energy. Our eyes are not sensitive to all the "blackbody radiation" determined by the temperature of the object. At ambient temperatures the radiation falls into various portions of the infrared spectrum. One class of sensor that is useful is a thermal energy imaging device. An example of one is shown in Fig. C-1. A more capable, sophisticated, and expensive device is shown in Fig. C-2.

An alternative class is one that makes use of low levels of light such as starlight, moonlight, and scattered light from nearby lighting. This example is shown in Fig. C-3. All military versions are expensive devices.

Because of civilian demand, much less expensive versions of these types of devices are becoming available. Catalogs for sportsmen, outdoorsmen, and the like regularly contain information for devices that can provide low light level visualization of terrain, and so on. These devices are typically ten to one hundred times cheaper than the military versions and are quite adequate for the general task of maintaining some kind of surveillance around a perimeter, outside a building, or along a line of movement. Exemplar sources are www.herters.com and www.skyandtelescope.com.

The last example is shown to provide a complete picture of the technology available, but it is probably an unlikely candidate. The advantage of the infrared searchlight (see Fig. C-4) is that cheaper or less expensive infrared systems combined with the searchlight could lower the overall cost of providing surveillance. It is not unusual for a full militarized infrared device to cost a few hundred thousand dollars.

The starlight telescopes available for civilian outdoor application are typically less than one thousand dollars. Clearly the militarized devices contain features that are not included in the civilian devices. They typically have large apertures, much lower detection thresholds, and other features that may or may not be useful but typically are needed for military operations.

The reasoning behind these examples can be seen in the smart buyer context. If a humanitarian operation is going to be conducted in an isolated setting, such devices could provide at least some measure of warning. Humanitarian operations that typically travel light and are lightly equipped except for humanitarian purposes would probably not actually purchase such equipment; however, the security forces that could support the humanitarian operations would have these. Knowing their performance would be useful in making arrangements for security.

Figure C-1

400D - Color Thermal Imaging

Category:
Night Vision and Optics

Equipment Name:
400D - Color Thermal Imaging
Point of Contact:
Name: Mr. Thomas Hurley
Phone: (410) 875-0234
Fax: (410) 875-0291

Email: lwhurley@hurleyir.com
Website: http://www.hurleyir.com

Company Information:
Hurley and Associates, Inc
P.O. Box 77
4757 Buffalo Road
Mt. Airy, MD 21771
USA

Description:

The Pro 400 Digital handheld color-imaging camera provides digital data storage and robust performance, all combined with Raytheon reliability. The Pro 400 Digital boasts an affordable price and excellent image quality.

With the power and flexibility of the industry's leading PDA, the Pro 400 Digital can store and recall over 150 images. Perfect for industrial or public safety applications, the Pro 400D can be used for predictive maintenance, process control, insulation analysis and general law enforcement surveillance operations.

Cost Data: Contact us for a price quote.

Figure C-2.

Multi-Sensor Surveillance System
with Model SPS-1000MS.

Category:
Night Vision and Optics

Equipment Name:
Multi-Sensor Surveillance System with Model SPS-1000MS.
Point of Contact:
Name: Mr. Bruce W. Harting
Phone: (727) 299-0150
Fax: (727) 299-0804

Email: bharting@xybion.com

Company Information:
Atlantic Positioning Systems
11528 53rd Street North
Clearwater , FL 33760
USA

Description:
The Atlantic Positioning System Multi-Sensor Surveillance System is a gimballed set of electro-optical sensors to include FLIR, LLTV, day video, and laser ranging/illumination devices. These sensors are selected for the user's missions, usually long-range surveillance, to provide detection, recognition, and identification of selected targets. Since the positioner is a precision device, it provides low jitter, clear images on fixed, mobile, marine, or airborne platforms. Depending on the size of the sensors, the system can be man portable, remotely operated, and operate from 24 VDC or 115 VAC power.

The systems are in production and used by the US Navy, UK MOD, ROC Navy, and other US agencies in fixed and stabilized modes. The systems can be controlled via personal computer, host processor, or APS's operator control panel.

Cost Data: The SPS-1000MS cost is highly variable due to sensor selection, modes of operation, precision, stabilization, and options. A typical system can vary from $100,000 to $700,000. Delivery is normally 6–12 months ARO.

Figure C-3.

170mm Long-Range
Night-Vision System

Category:
Night Vision and Optics

Equipment Name:
170mm Long-Range Night-Vision System
Point of Contact:
Name: Mr. Mike Matzko
Phone: (724) 295-2880
Fax: (724) 295-2336

Email: oststar@nauticom.net

Company Information:
Optical Systems Technology, Inc.
110 Kountz Lane
Freeport , PA 16229-1724
USA

Description:

The OSTI(Star-Tron)Long-Range 170mm fast f/1.4 catadioptric night vision lens offers 6.5X magnification in the most demanding low light level applications. The optics are coated for visible and near IR light, thus allowing the current Gen-3 image intensifiers to perform at maximum. The lens will adapt to a variety of NVDs using 18mm or 25mm image tubes. Viewing ranges with Gen-3 intensifiers at Full Moon Illumination are 6,000 meters and at Starlight Illumination 2,000 meters.

Cost Data: Contact us for a price quote.

Figure C-4

Dragon Night Star
Infrared Search Light

Category:
Night Vision and Optics

Equipment Name:
Dragon Night Star Infrared Seach Light
Point of Contact:
Name: Mr. William Conklin
Phone: (831) 728-9090
Fax: (831) 728-1964

Email: b@lifesafetysys.com
Website: http://www.lifesafetysys.com

Company Information:
Life Safety Systems
343 Soquel Avenue
Suite 317
Santa Cruz, CA 95062
USA

Description:

50 watt, 800,000 candlepower High Intensity Search light system with infrared filters for nighttime, night vision surveillance use. Provides multiple IR filter densities for long range night vision illumination. System comes complete with Dragon Star light, 3- IR filters, 110 vac charger, 12vdc power cord and heavy duty storage case.

Cost Data: Contact us for a price quote.

IMPROVEMENTS FOR FACILITY PHYSICAL PROTECTION

The next three exhibit sets encompass a variety of protective means for facilities and the people within. They range from barriers to blankets to fire retardant coatings and panels.

Once again, the items shown should not be considered as those which might be purchased by humanitarian organizations except in unusual circumstances. The issue here is what should be demanded for protection from local governments that often have such equipment.

D. Barriers

Protecting buildings and the people within starts at perimeters displaced from the facilities. Physical barriers of two types are shown, although there are other types.

The first is an architectural barrier, shown in Fig. D-1. This type incorporates space for aesthetics that may already be part of the facilities used for humanitarian work and workers. The cruder

Figure D-1

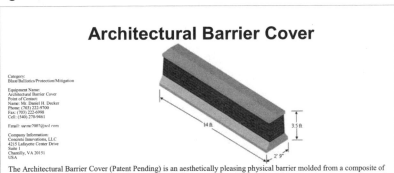

Architectural Barrier Cover

Category:
Blast/Ballistics/Protection/Mitigation

Equipment Name:
Architectural Barrier Cover
Point of Contact:
Name: Mr. Daniel H. Decker
Phone: (703) 222-9700
Fax: (703) 222-6998
Cell: (540) 270-9461

Email: ssenc2002@aol.com

Company Information:
Concrete Innovations, LLC
4215 Lafayette Center Drive
Suite 1
Chantilly, VA 20151
USA

14 ft   3.5 ft   2' 9"

The Architectural Barrier Cover (Patent Pending) is an aesthetically pleasing physical barrier molded from a composite of concrete and other materials. The exterior can be molded with a variety of decorative finishes to enhance or complement the surrounding architectural features. The interior is open to accommodate a standard roadway barrier, cast in place concrete, absorbcrete, soil, or other materials. The Architectural Barrier cover can be opened through the top to accommodate plantings or capped with a decorative cover.

Functionality: The Architectural Barrier Cover is an adaptable alternative to standard roadway barrier or attractive cover to slip over an existing barrier for use in a perimeter security plan. The cover can be used to establish a stand-off perimeter, discourage vehicle intrusion, and funnel traffic through secure access points. The barrier cover is portable and can be moved to adapt to changing threats and traffic patterns. The Architectural Barrier Cover enhances the crash resistant qualities of an already existing barrier and, by adding a variety of patented materials and structural components, decreases vehicle penetration significantly.

Cost Data: The Architectural Barrier Cover is custom manufactured with a variety of exterior finishes and interior configurations. The basic configuration is priced at $1,285 each, delivered in our local area. Government and quantity discounts are available. Please contact our Product Manager, Anthony Wambaugh, P.E., (703) 222-9701, ext 154.

form of these are so-called Jersey walls used in many security and safety applications. The robust versions of these can be used to establish both access routes and perimeters that limit facility areas and levels of damage from, for instance, truck bombs.

In more sophisticated settings, movable access barriers shown in Fig. D-2 can both regulate traffic—human and vehicular—and provide crash resistance for additional protections.

Figure D-2

Automatic Barrier Gate
& Barrier Crash Gate

Category:
Delay and Denial Technology Including Barriers

Equipment Name:
Automatic Barrier Gate & Barrier Crash Gate
Point of Contact:
Name: Mr. Max Krake
Phone: (225) 274-1115
Fax: (800) 676-5535
Toll-Free: (800) 676-5537

Email: max@ldi.com
Website: http://www.ldi.com

Company Information:
Logical Decisions Inc.
2020 N Sherwood Forest Boulevard
Baton Rouge, LA 70815-1959
USA

Description:
LDI-004CR -- crash rated at K4/L3.

Off-the-shelf item with typical 35 days delivery. Barrier arm will cover a 16' to 20' curb-to-curb opening. Open or close in 8 seconds and contains two 1' steel cables, each with a breaking point greater than 100,000 lbs. All systems use the world's number one hydraulic gate operator UL approved for class 1 through 4. Maintenance is near non-existent. Speed is on the high end and strength is superb. Each gate is designed to fit the exact need.

Cost Data:
$17,000 to $20,000

E. Protective Blankets

Protective blankets come in a large variety of forms for an equally large set of functions and applications. The threats considered range from explosives, which primarily generate blast environments and secondary debris environments, to blast, fragmentation, and slug and shaped charge munitions.

Blast suppression and containment blankets are shown in Fig. E-1. As indicated in the accompanying text, such blankets have resisted explosive environments ranging from 4 lb. to 250 lb. TNT charges.

Figure E-1.

Blastnet - Blast Suppression

Category:
Blast/Ballistics/Protection/Mitigation

Equipment Name:
Blastnet - Blast Suppression
Point of Contact:
Name: Mr. Dan Gallucci
Phone: (770) 844-9438
Fax: (770) 844-9438

Email: dan@millibar.com
Website: http://www.millibar.com

Company Information:
New Necessities
5710 Pebble Brook Trail
Gainesville, GA 30506
USA

Description:

Unique thin, soft, flexible advanced materials can be incorporated into virtually any vulnerable area. Specifications are totally dependent on threat assessment and site requirements, but can be manufactured in nearly any size or shape; can be built-in or temporary (high threat levels), restrained, or draped; and has patented ported, over-pressure reduction and sacrificial delamination features. Highly efficient blast suppression; debris capture; over-pressure reduction; soft yield; and progressive failure.

At FPED III, Blastnet was tested with 4 lbs. of C-4, 5' standoff, with debris field generators (computer/office equipment), RESULT: No penetration and no failure; the computer behind Blastnet was COMPLETELY UNHARMED (left half of image). In separate blast tests in conjunction with DoD and other agencies, Blastnet successfully withstood explosions of up to 250 lbs. of TNT (right side of image).

Cost Data: Contact us for a price quote.

Figures E-2, 3 and 4 show blankets that provide resistance to blast and fragmentation, including smaller slugs. The differences lie in applications. The blanket type shown in Fig. E-2 is typically draped over the charge or munitions. Those in Figs. E-3 and E-4 are configured to be used as "temporary walls"; the former is designed with a ladder, and the latter is attachable in some way.

While the example applications have a facility setting, such blankets can be used to protect vehicles, stores, or important infrastructures such as generators. As such, they must be adapted to operating conditions. Cooling, for example, must be provided.

F. Fire Retardant Panels

Two examples shown in Figs. F-1 and F-2 take different approaches to protecting vital elements of facilities such as the portions of buildings that might be regarded as safe rooms and would only be used where there is some expectation of permanence in these facilities. The properties of these fire retardant technologies are fully described in the figures and are not repeated here.

Figure E-2.

Ballistic Blankets, Bullet Resistant and Fragmentation

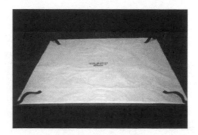

Category:
Individual Protective Equipment

Equipment Name:
Ballistic Blankets, Bullet Resistant and Fragmentation
Point of Contact:
Name: Mr. Thomas G. Faust
Phone: (610) 375-8549
Fax: (610) 375-4488

Email: bulletpr@bellatlantic.net

Company Information:
T. G. Faust Inc.
544 Minor Street
Reading, PA 19602-2722
USA

Description:

There are two types of blankets offered: 1) BOMB BLANKETS Description: The blanket is effective against most pipe bombs, hand grenade fragments and most fragmentation caused by pressure or electrical explosion. Construction: A combination of layers of ballistic fabric sewn together in a pattern to produce maximum strength then sewn into a flame, acid and water repellant cover. 2) BARRIER BLANKET Description: This bullet-resistant unit is available in ballistic protection Levels IIA, II or IIIA. Construction: Produced from multiple layers of Aramid fabric conforming to the protection level required, then sewn into a water repellant nylon cover. All blankets are available in the following sizes: 4'x4', 4'x5', 4'x 6', 5'x 6', 6'x 6' Custom sizes or configurations are available on request.

Cost Data: Contact us for a price quote.

Figure E-3.

Portal Blanket – Ladder / Ballistic Shield Combination

Category:
Blast/Ballistics/Protection/Mitigation

Equipment Name:
Portal Blanket - Ladder/Ballistic Shield Combination
Point of Contact:
Name: Mr. Don R. Budke
Phone: (513) 742-7100
Fax: (513) 853-3605
Toll-Free: (800) 346-6699

Email: dbudke@reliancearmor.com

Company Information:
Reliance Armor Systems
3107 Spring Grove Avenue
Cincinnati, OH 45225
USA

Description:
Provides level IIIA protection
Integrates with PORTAL Ladder (included)
Use for cover for high risk approach
Provides blast protection for breaching
Converts to a portable stretcher

Washable cover
Capable of being repelled vertically
Separate and use as a bomb blanket
Full deployment in 1 second!
Folds for compact storage in 5 seconds!
Self-storage when separated from ladder
100% waterproof, resists mold and mildew!

Cost Data: Contact us for a price quote.

Figure E-4.

Barricade Breaching Blanket

Category:
Individual Protective Equipment

Equipment Name:
Barricade Breaching Blanket
Point of Contact:
Name: Ms. Lisa Sagal
Phone: (954) 630-0900 Ext: 307
Fax: (954) 334-1702
Toll-Free: (800) 413-5155 Ext: 307

Email: lsagal@pointblankarmor.com
Website: http://www.pointblankarmor.com

Company Information:
Point Blank Body Armor
4031 NE 12 Terrace
Oakland, FL 33334
USA

Description:
Point Blank's Barricade Breaching Blanket is intelligently designed to provide a protective barrier for personnel against Level II or Level IIIA ammunition, while reducing the risk of injury due to fragmentation and other related effects from munition detonations.

Cost Data:
32"x72", NIJ Level IIIA - $3000.00
36"x72", NIJ Level IIIA - $3500.00
48"x72", NIJ Level IIIA - $5000.00

Figure F-1.

CKC Intumescent
Fire Retardant Coating

Category:
Individual Protective Equipment

Equipment Name:
CKC Intumescent Fire Retardant Coating
Point of Contact:
Name: Ms. Clariece Brecht
Phone: (320) 563-0116
Fax: (320) 563-8299

Email: geo@traversenet.com

Company Information:
Paragon Coating Services, Inc.
Route 1 Box 141
Wheaton, MN 56296
USA

CKC-268 standard 1 gal. Package.   CKC-268 result after 10 min.
5 gal. Package avn.   Exposure to flame

Description:
Interior CLASS A, water based, non-toxic, flat, intumescent flame-retardant coating
UL 723
Test for Surface Burning Characteristics of Building Materials R20412
ASTM E-84-99
Flame Spread Index 5
Smoke Developed Index 20

Cost Data: Contact us for a price quote.

Figure F-2.

Accordion Type Fire Folding Partitions (Fireguard)

Category:
Physical Security Equipment

Equipment Name:
Doors: Accordion Type Fire Folding Partitions (Fireguard)
Point of Contact:
Name: Ms. Kathy Wright
Phone: (410) 552-9950
Fax: (410) 552-9939

Email: kwright@govsupply.com
Website: http://www.govsupply.com

Company Information:
GovSupply (A division of Intellimar, Inc.)
7566 Main Street
Suite 113
Sykesville, MD 21784
USA

Description:
GovSupply supplies the Won-Door FireGuard product line. This unique series of high speed computerized accordion doors provides both fire safety and access control components within a facility.
These doors are being used extensively in key military facilities in both new construction and after-market applications to meet both code requirements and also provide the ability to section off a building the event of a security breach. These doors are the newest line of defense in facility interior entry control solutions. When all other security measures have failed, security personnel can immediately activate this physical access control measure to break the facility into predefined sectors and address the threat.
All GovSupply products are available through GSA Schedule# GS-07F-0100M (Intellimar, Inc.).

Cost Data: Contact GovSupply at (410)552-9950 for GSA Schedule pricing [Contract #GS-07F-0100M]

These facility-related examples could provide the humanitarian organizations with discrimination criteria to use in selecting from facilities offered to them. They also provide information to help in specifying to host governments or groups what might be needed in the way of fire protection.

G. Unmanned Aerial Vehicles

Three examples of unmanned vehicles are presented for discussion and consideration. They include two very small aircraft and an aerostat, shown in Figs. G-1, 2 and 3. How might these be used?

The aerostat would provide an elevated, inexpensive platform that would definitely improve communications in areas where wireless communications are needed. Often the terrain presents real obstacles for propagation. Mountains, vegetation, and swamps all make propagation quite difficult, and as a result organizational elements are out of communications with one another.

Having an elevated antenna supporting such communications improves the circumstances markedly. The aerostat example cho-

Figure G-1.

Mobile Observation Aerostat

Category: Unmanned Aerial Vehicles

Equipment Name:
Mobile Observation Aerostat
Point of Contact:
Name: Ms. Noga Rosenblum
Phone: (301) 913-9369
Fax: (301) 913-9366

Email: bresweber@mistralgroup.com

Company Information:
Mistral Security Inc.
7910 Woodmont Avenue
Suite 820
Bethesda, MD 20816
USA

Description: The mobile observation aerostat was specifically designed to provide a long endurance, high vantage point observation platform that is fully compatible with urban environments.
Aerostat: Most effective in areas where the geography prohibits the use of more conventional surveillance capabilities or where a temporary "look down" capability may be required. The aerostat is 26' long and is carried inflated in a trailer towed by a commercial truck or SUV.
Characteristics:
- Deployed in under 20 minutes, by 3 persons.
- Operating altitude up to 500'.
- Can detect a human at a range of 2.4 miles.
- Capable of day/night observation from a 3 axis stabilized payload.
- Non-permanent, no environmental impact issues.

Cost Data: Contact us for a price quote

Figure G-2.

Miniature UAV

Category:
Unmanned Aerial Vehicles
Equipment Name:
Miniature UAV
Point of Contact:
Name: Ms. Noga Rosenblum
Phone: (301) 913-9369
Fax: (301) 913-9366
Email: bresweber@mistralgroup.com
Company Information:
Mistral Security Inc.
7910 Woodmont Avenue
Suite 820
Bethesda, MD 20816
USA

Description:
Lightweight, low cost aerial observation platform. Conducts reconnaissance and surveillance of aircraft approach and departure paths, out to a distance of 6 miles. "Real Time" down link video couples to the on-board GPS; compass and altimeter provide real time information and location data for ground security force reaction, if an anomaly is detected.
Characteristics:
- Hand launched / 2-man team
- Manual operation / Preprogrammed flight patterns
- Utilizes digital maps / Auto-return to "home" if signal is lost
- Powered by electric motor / no environmental noise impact
- Full day / night sensor capability
- Up to 60 minutes flight time
- Operational in winds up to 25 Knots [30 MPH]
Cost Data: Contact us for a price quote

Figure G-3.

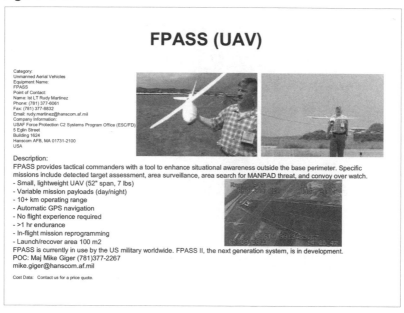

sen is not inexpensive, but there are inexpensive balloons that could be used for such purposes. These are used by weather services, for instance, and have long endurance, particularly when tethered.

The other alternative would be small unmanned aircraft. Surprisingly, these are inexpensive, but they require training and some ability to maintain. Three applications come to mind. The first is general surveillance around a compound, particularly in times when violence is expected. The ability to use the simple optical devices (equivalent of a TV camera) in such unmanned aerial vehicles provides an extension of situation awareness without putting people on the ground around perimeters.

The second use is providing some situation awareness on transit routes for movement through terrain suspected to be or at least anticipated to be problematic. The third mission for such machines, especially the somewhat larger ones, is to carry important supplies from one place to another, particularly in rugged terrain.

Unmanned aerial vehicles are much cheaper than aircraft and

often require much in the way of open space for takeoff and landing. Some of these have payloads ranging from ten to one hundred pounds, and the speeds at which they operate are on the order of typically one hundred miles an hour.

Using unmanned aerial vehicles, the movement of critical cargo, medical supplies, for example, could be accomplished in a matter of hours without driving or portering through hostile or rugged terrain or a combination thereof. This is an unusual way of looking at these vehicles, but it has its possibilities.

H. Security Assessment Software

Software is now available to assist in a variety of facility assessments, movement, and other activities that stress the safety of humanitarian workers. The examples shown include site, contingency, and crisis information management tools. There are software packages that have been employed by the military and by organizations involved with consequence management such as the National Guard and emergency first responders.

The three examples shown are:

Figure H-1 (Site Profiler)
Figure H-2 (EIS/GEMN Contingency Management Software)
Figure H-3 (CIMS: Crisis Information Management Software—Feature Comparisons Report)

These kinds of tools are considered necessary in combination for a pre-assessment and planning phase activity that addresses security and safety. The results with on-site refinements should be employed in training and in operations.

The National Institutes of Justice (NIJ) Crisis Information Management Software—Feature Comparison Report referred to on Fig. H-3 was published in October 2002 and can expedite selecting software for specific sites or contexts. It is available for free download at *http://www.ojp.usdoj.gov/nij/pubs-sum/197065.htm.*

At this point one must address databases that are required to get useful results from the software tools. Geographic and map data could be made available by host governments or deduced from satellite imagery now available from global sources. Site and building information must be gathered from host sources. In the

Figure H-1.

Site Profiler Assessor
(Office of Domestic Preparedness Edition)

Category:
Vulnerability Assessment Software

Equipment Name:
Site Profiler Assessor (Office of Domestic Preparedness Edition)
Point of Contact:
Name: Mr. Bryan S. Ware
Phone: (703) 871-5102
Fax: (703) 871-5103
Cell: (703) 981-4033

Email: bware@dsbox.com
Website: http://www.dsbox.com

Company Information:
Digital Sandbox, Inc.
11710 Plaza America Drive
Suite 2000
Reston, VA 20190
USA

Description:
The Office of Domestic Preparedness (ODP) edition of Site Profiler Assessor, is a special version of Site Profiler Assessor for State and Local governments. The Site Profiler Assessor ODP Edition facilitates a Vulnerability Assessment in accordance with standards published by the Department of Homeland Security's Office of Domestic Preparedness.

Cost Data:
$1,500 - 2,500 (depending on configuration and number of licenses purchased)

Figure H-2.

EIS/GEM v8.3
Contingency Management Software

Category:
Vulnerability Assessment Software

Equipment Name:
EIS/GEM v8.3 Contingency Management Software
Point of Contact:
Name: Ms. Maxine Orens
Phone: (301) 556-1728
Fax: (301) 556-1701
Toll-Free: (800) 999-5009 Ext: 1728
Cell: (301) 509-7180

Email: maxine_orens@environ.com
Website: http://www.essential-technologies.com

Company Information:
Essential Information Systems, Inc.
1395 Piccard Drive
Suite 230
Rockville, MD 20850
USA

Description:
EIS/GEM v8.3 is a contingency management software solution for incident response, daily operations, planning, exercising, and training. It brings together maps, models, data, and communications to handle any type of natural and technological disaster planning and the command and control of responses to such incidents.

Cost Data: Available on the GSA Schedule.

Figure H-3.

Crisis Information Management Software (CIMS)

Category:
Vulnerability Assessment Software

Equipment Name:
Crisis Information Management Software (CIMS)
Point of Contact:
Name: Ms. Julie Anderson
Phone: (703) 465-4600
Fax: (703) 243-2047

Email: janderson@camber.com

Company Information:
Camber Corporation
3855 Center View Drive
Suite 400
Chantilly, VA 20151
USA

Excellence Through Teamwork

Description:
The National Institute of Justice (NIJ) Special Report compares 10 Crisis Information Management
Software (CIMS) products used by emergency management agencies (EMAs). The products examined
were specifically designed to augment EMA responses to crisis situations and enhance emergency
management planning and mitigation. This report is accompanied by the interactive Excel-based CIMS
Feature Comparison Matrix that allows an agency to assess the performance of a product(s)—respective
of agency priorities, requirements, and conditions of operation—and use that information for the
procurement process.

Cost Data:
Contact us for a price quote.
No cost to obtain CIMS Feature Comparison Report or Matrix.
Please visit the following website to download copies.
http://www.ojp.usdoj.gov/nij/pubs-sum/197065.htm

worst circumstances it will come from the deployed teams, although it is clearly a burden that will interfere with the delivery of humanitarian services.

All is not lost even if such databases are not available for impending operations. The training value of the tools and methodologies is still available using historical examples. This latter point will be discussed in greater detail in the next section.

I. War-Gaming and Simulation for Training

While the majority of this chapter is focused on technology, training is a vital enabler. Today three different sets of tools are used for training in the most advanced circumstances by the military, emergency workers, and those involved in dealing with consequence management of small and large incidents ranging from accidents to disasters. The relationship between simulation and training might best be shown with examples from other domains.

The Department of Defense and other departments of the U.S. government train people in three different environments and sometimes combine the environments. The environments are characterized as live, virtual, and constructive. Combinations of these are also employed using networks. This has the advantage of providing excellent training for units and people who are geographically dispersed. Some of these aspects could be important even for training humanitarian organizations.

Live training is accomplished using an instrumented facility. With this facility the actions, decisions, and movements of all individuals and platforms and (weapons) are accounted for. The largest of these facilities are located in the western part of the United States at the National Training Center, Nellis Air Force Base, Fallon Naval Air Station, Twenty-Nine Palms, and the Navy Top Gun facility. The instrumentation provides what is called ground truth so there can be no dispute as to causes and effects, actions and outcomes.

Based at these facilities are so-called opposing forces. These units are trained to operate as expected opponents might operate. In addition, they are encouraged to innovate so that reactive enemies and asymmetric threats are inflicted on the groups of people being trained.

The methodology described thus far was initiated in the 1970s and dramatically changed the capabilities of our military forces. Similar training regimens are employed by the Department of Justice, Department of Energy, and state and local governments in training their first responders, firefighters, and police. The simple description of the live training is that the circumstances in effect put the trainees against a world-class opposed force. In combat circumstances in recent times, some participants have actually said that the training was more difficult than the military operations against real foes.

Virtual simulation is employed to immerse the trainees in an environment that is sufficiently realistic such that they experience all or nearly all of the circumstances they would encounter in live settings. A commercial example of the virtual training environment is the one used to train pilots and crews for commercial aircraft. These training exercises use simulators that move with six degrees freedom, and the stimuli for the trainees is provided by visual characterizations of different circumstances driven by computers. The same computers drive the instruments on the aircraft. In effect, the pilots and crews are immersed in a virtual environment that is very realistic. In that environment, they are subjected to emergencies that would be life threatening if they occurred in a live environment.

The third simulation mode employed in such training is called constructive. It ranges from games on tabletops (participants sitting on opposite sides of the table) to computer-based war games to multiplayer war games. The advantage lies in two areas for such simulations employed for training. The first involves cost, as these types of training are quite inexpensive if one sets aside the manpower cost. The second advantage lies in the ability to make many repetitions of the circumstances with variations in them. Constructive simulations are employed to understand the variations and sensitivities to changes not in the absolute outcomes.

Today's most sophisticated computer games are referred to as massive multiplayer games. In these there is actually a crossover from simulations for training to the entertainment world. This point will be discussed further in this section.

Considering these attributes, such tools may be useful for recruiting and should be useful for training and for planning and structuring organizations. How expensive are they? The answer to that question depends upon the tools.

The massive multiplayer games that involve hundreds of thousands of people on-line (who pay $10 or so a month to play) cost anywhere from $20 to $40 million. This clearly is not something that an NGO would undertake by itself. However, these are products of the entertainment industry. For the most expensive games at least, wealthy contributors from the entertainment industry might be willing to provide in-kind products instead of or in addition to money.

The lesser forms of these simulations are relatively inexpensive. The tools described in the previous section fall in this category.

The final suggestion made here is that the groups of humanitarian organizations that provide most of the services in the world might approach a government sponsor. Since they work very closely with state and defense, these departments might be convinced to undertake development and maintenance on behalf of the humanitarian organizations.

What does this all mean for humanitarian workers, and how

might this technology be leveraged? In the first instance, the training that can be achieved is superior to simply mentoring, depending upon exactly the mix of modes and the role of the would-be mentors. In many of the previously cited circumstances, mentoring is done by a separate group of people, and the mentoring is constructive not critical.

With such tools, it is possible to immerse people in circumstances they may encounter, and therefore they are more likely to respond correctly, having experienced the challenges in their training environment. Such methods are also useful in screening people who might be recruits.

Figures I-1 and I-2 are included in this section. They are drawn from a web-based simulation called America's Army, which is used as a recruiting tool and has been very effective. Within it are elements of the virtual and constructive simulation characteristics were described earlier. The first element displayed is simply the top-level descriptor of America's Army.

The second, however, has to do with markets and in that regard may provide an opportunity for humanitarian organizations. The

Figure I-1.

Figure I-2.

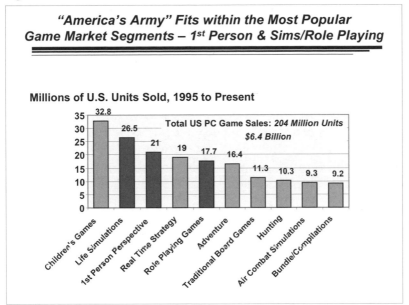

"America's Army" Fits within the Most Popular Game Market Segments – 1st Person & Sims/Role Playing

Millions of U.S. Units Sold, 1995 to Present

Total US PC Game Sales: *204 Million Units*
$6.4 Billion

Children's Games: 32.8
Life Simulations: 26.5
1st Person Perspective: 21
Real Time Strategy: 19
Role Playing Games: 17.7
Adventure: 16.4
Traditional Board Games: 11.3
Hunting: 10.3
Air Combat Simulations: 9.3
Bundle/Compilations: 9.2

gaming market, which involves individual participation, is growing and growing rapidly. The design characteristics shown in the second chart provide an insight into the objectives. The army is attempting to portray a realistic army environment that provides recruits the opportunity to experience the army as a career as well as an individual challenge. It is a web-based simulation. By analogy, this approach would help humanitarian organizations with both recruiting and training.

CONCLUSION

The long litany of technology examples could illuminate ways in which safety and security can be enhanced. Ideas and examples well beyond those displayed here are surely worth exploring because of the inherent value of lives saved. It is clear from the symposium in which these and other possibilities were explored that there is a structural gap which is unfilled.

There is no sponsoring activity to bring useful technology into humanitarian operations other than that provided by individual

NGOs. None can afford to purchase more than off-the-shelf items. The ability to obtain items and services from host countries is limited by the general poverty of the peoples and governments served.

The only possibilities open are from either long-term commitment for resources obtained from a consortia of NGOs or help from departments of the U.S. government. The expertise for technology acquisition and development clearly lies in certain U.S. government departments—defense, energy, transportation, intelligence—but not in their missions. Providing improved security and safety will require mission expansions by these departments if technology is to play a role.

Humanitarian crises appear in the media as sudden and dramatic: images of starving children, floodwaters rising, refugees arriving at camps. But most crises have taken time to develop: more than one crop has failed, the rains were predicted, the conflict has intensified. Governments and major agencies have Early Warning Systems that begin to set off alarm bells.

Unfortunately it is easier to find funding for an emerging crisis than it is to find funds for the prevention or mitigation of an impending crisis.

On December 26, 2003, when most of the population was sleeping, an earthquake scoring 6.7 on the U.S. Geological Survey scale hit the two thousand-year-old city of Bam in Iran "burying 43,000 under the bricks, the rubble, and the mud and destroying the two public hospitals in the city. Nearly half of the ancient city's inhabitants died," according to the spokesperson from the International Federation of Red Cross and Red Crescent Societies, which deployed 7,000 relief workers and provided "90,000 tents, 200,000 blankets, 56,000 pieces of clothing, and 51,000 kerosene heaters to the 70,000 homeless."

All of this chaos and tragedy occurred in a country that has experienced many earthquakes. Journalists and analysts were quick to ask "Why". What had failed? Were there no early warning indicators? Had the architects of the buildings neglected to consider the prevailing natural conditions? Had builders ignored or adjusted tolerances set by the architects?

Can future earthquakes, volcano eruptions, floods, famines, droughts, and other natural disasters be avoided or mitigated by preemptive knowledge?

"The civil war that erupted in Rwanda in April 1994 claimed more than 800,000 lives and forced more than a third of the country's 7.3 million people from their homes. More than 2.1 million Rwandese sought asylum in neighboring countries."

This stark statement from UNHCR can be better appreciated viewed on a smaller scale. "The Rwandan genocide was in full swing and there were rumours that a refugee exodus of biblical proportions was under way," wrote Ray Wilkinson of UNHCR.

In Tanzania Maureen Connelly and other aid workers anxiously visited the border area almost daily to check, but at first there was only an ominous silence. "There was no movement," recalls Connelly. "There was no information. Had the genocide swallowed these people up as well? Did they even exist?" A few days later on April 28, "We looked up at the Rwandan hills. There was nothing but people. The hills were covered with a moving mass. The entire African landscape was awash with people all headed our way." Wilkinson, in his article for *Refugees* magazine, added that 200,000 Rwandans crossed in twenty-four hours into Tanzania through this single border post. "Hundreds of thousands of people fled in every direction of the compass in 1994."

Could all this have been seen from the sky, sensed from the ground?

—Larry Hollingworth

High-Resolution Earth Monitoring for Humanitarian Action

Arthur Lerner-Lam, Ph.D., Kristina Rodriguez Czuchlewski, and Jeffrey G. Weissel

TECHNOLOGIES for earth observation have advanced significantly in recent years, driven principally by advances in space and airborne sensors. New instrumentation deployments now provide unprecedented views of Earth's surface across a broad spectrum of wavelengths and resolution capabilities. Most applications of these new technologies are for scientific research, national security, or commercial purposes. However, the capabilities inherent in these observations systems should give rise to new opportunities for their use in humanitarian intervention and sustainable development. This chapter provides a brief review of earth observation capabilities within the framework of environmental stress and natural disasters, and suggests areas of research and development that could make the information products more useful to humanitarian workers.

INTRODUCTION

While not all humanitarian action is precipitated by natural disasters, the frequency of disasters, their disproportionate impact on the poor, and their occurrence without regard or prejudice to cultural or political environment make disaster observation and emergency response a proving ground for the use of new technologies to support humanitarian action. This is the case particularly for new technologies in high-resolution earth observation. Natu-

This chapter is catalogued as Lamont-Doherty Earth Observatory Contribution Number 6634.

ral disasters create the need for humanitarian intervention in the form of rapid emergency response. However, new predictive capabilities in earth science, especially those that rely on high-resolution earth monitoring, raise the possibility that humanitarian actions can be used before the occurrence of a disaster to mitigate its impacts. Such an enlarged framework for humanitarian action serves a dual purpose: by building disaster resiliency, humanitarian action can contribute to a more sustainable and civil society. Both pre-event and post-event instances of disaster-driven humanitarian action rely on accurate and rapid characterization of natural disasters and their impacts on human settlements. Thus there is a direct link between technology and humanitarian action.

The impacts of disasters on humanity have many dimensions, including the loss of life, loss of individual livelihood, and the reduction of economic performance and social function. Disasters usually affect the poor more than the rich, and poverty often amplifies the immediate as well as long-term impacts of natural stress. Each year, considerable numbers of persons are rendered homeless or become refugees as a direct consequence of disasters. Additionally, economic and political events that by themselves provoke humanitarian crises can be triggered or amplified by disasters that exceed the coping capacities of marginal states. Inasmuch as disasters and their impacts are observable events, earth monitoring and observation should play a significant role in the preparation and implementation of pre-event humanitarian intervention or post-event response.

The technology of earth observation is broadly classified as either *remote* or *in situ* sensing, and both are relevant to humanitarian action. In addition, the data provided by earth observation systems is not generally useful unless preliminary analysis and quality control are completed and data products serving particular stakeholder communities are produced. Thus, humanitarian action is linked to earth observation information management, not just raw data.

In this brief summary, we describe general features of earth observation technologies and describe their importance to predictive earth science. We then list some issues associated with data use for humanitarian action and conclude with some recommendations.

SUMMARY OF EARTH OBSERVATION TECHNOLOGIES

In Situ and Remote Sensing

"Earth observation" comprises hardware (sensors and platforms) and software (data analysis, management and distribution systems) that serve to characterize the Earth's surface and the near surface environment. Earth observation systems generally are of two types: *in situ* and *remote*. *In situ* systems normally occupy the space they are observing and make so-called "point" measurements or measurements in the vicinity of the sensor deployment. Often, arrays or networks of in situ sensors can be deployed to increase the geographic coverage of observation. Because in situ sensors are closely coupled with the environment they are observing, we often say that they provide "ground truth."

While useful, in situ sensors are also prone to environmental contamination: noise processes can corrupt the desired signal and limit its utility for a particular application. To deal with this, in situ sensors are often deployed in spatially dense arrays or over longer periods of time in order to develop observational redundancy that may reduce the effects of undesired noise.

In contrast, remote sensing systems observe the Earth's surface from some distance, relying on reflected or transmitted electromagnetic, optical, or acoustic radiation to probe the area of interest. Sensors can be flown on aircraft or on orbiting platforms, or they may be deployed from ships at sea or vehicles on land. Fundamentally, remote sensing requires that something be known about the propagation of the sensing radiation through the intervening medium; the deterioration of the radiation as it propagates, and shielding by obstructions such as clouds, ground cover, or unknown structure, degrades the image. The deployments of remote sensing systems are designed with these limitations in mind: observations of different phenomena require different types of sensors measuring different types of energy. Like in situ systems, remote sensing measurements are also subject to contamination by noise: most systems employ special data processing algorithms or deployment strategies to reduce noise problems. Where possible, in situ sensors are often used to calibrate the signals observed by remote observation systems, thus providing ground truth.

Both types of earth observation are important. First, in situ and remote sensing may not be capable of observing the same phenomenon. For example, the offset of faults during an earthquake may be observable from space, but the resonances of damaging earthquake waves are most accurately recorded by in situ instruments. Second, even if in situ and remote sensors target the same phenomenon, they cover different space and time scales. Remote sensing technologies generally are not capable of continuous monitoring, especially at high resolution. On the other hand, it is much easier to have broader geographic coverage with remote sensing.

Consequently, effective use of earth observations requires complementary information from a variety of platforms and from both in situ and remote systems. Integration of data from various systems is the key to effective support for humanitarian action.

The Interactions of Scale

The earth's surface is a manifestation of natural processes that arise from sources within the planet, in the oceans, and in the atmosphere. Human activity can modify the surface, of course, through either physical activity such as agriculture or urbanization, or through less visible but nevertheless detectable chemical or thermal contamination of the land, air, and water. Both human and natural processes operate at intrinsic and varied spatial and temporal scales, and it is the intersection of these processes across scales that limits the utility of any single earth observation system. Thus the information needed for humanitarian action must come from a collection of observations. As we discuss later, the information management system, used for decision support, should be the integrator.

Thus, scale provides the framework for discussing earth observing technologies and the information management systems that turn their raw output into useful products. Figure 1 shows a summary of spatial resolution and image-repeat times for representative satellite remote sensing systems. The spatial resolution, that is, the minimum length dimension that can be unambiguously observed (or, alternatively, the minimum separation distance that allows the differentiation of adjacent features), depends on both the sensor technology and the flight characteristics of its carrying platform. The observation repeat time, that is, the ability to observe the same location at two different times, is related to the aperture or "footprint" of the sensor on the surface and the orbital parameters of the platform. In general, "good" resolution and "rapid" repeat times are inversely related.

As shown in Fig. 1, satellite systems capable of resolution on the order of tens of meters have orbital repeat times of weeks to months, effectively reducing their efficacy for real-time observa-

Figure 1.

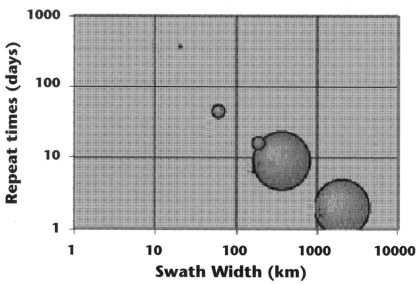

Data from Gubbels and Weissel (2002). Repeat times plotted against swath width for a selection of widely available government and commercial satellite remote sensing systems. Performance characteristics are plotted in log-log space to emphasize the inverse power-law relationship between the "footprint" of the sensing instrument and the ability of the platform to provide a repeat pass. The size of the symbols is related to the nominal spatial resolution length provided by the instrument. Thus sensors with large footprints have reduced resolution.

tions of small-scale phenomena. However, these nominal space and time resolution limits are based on what could be termed "first-order" processing. In fact, the physical signals that comprise remote sensing applications contain more information than is commonly realized. Part of the basic research agenda is the extraction of more information from a higher-order analysis of the remote sensing signals.

The antagonistic relationship between spatial resolution and temporal resolution is shown schematically in Fig. 2. Although new technology is always being developed, there will continue to be a trade-off between higher spatial resolution and rapid repeat observations. This is the "resolution trade-off." Advances in sensor technology and new platform capabilities, such as those afforded by autonomous unmanned aircraft, may permit rapid repeat or near-continuous observation in localized regions. However, this strategy requires new mission definitions and may not work in a pre-event mode, being practical only in post-event or emergency response situations.

Optimizing the resolution trade-off is a complex problem. On one hand, the trade-off is a consequence of the economic and logistic costs of deployment of individual technologies. On the other hand, there are real physics-based limitations on the utility and analytical capacity of data in some humanitarian applications. Most importantly, the trade-off between small-scale spatial observation and the need for real-time monitoring is a dominant constraint on the range of products available to humanitarian end users. In real applications, the trend toward higher spatial resolution may saturate at a level below the needs of humanitarian applications. A similar saturation limit may not hold for improvements in repeat time, however. While improvements in time-resolution aid phenomenological characterization in common with scientific modeling capabilities, direct observation in real time, with near-zero latency in transmission to the end user, has axiomatic utility in humanitarian applications.

Figures 1 and 2 illustrate a central dilemma of remote sensing: high-resolution monitoring must be targeted. It may be difficult to task the appropriate platform for a particular locality when disaster strikes. The utility of systems for humanitarian action rests on the ability of stakeholders to combine pre-event baseline information obtained from global or regional monitoring with real-time observations obtained from more flexible or in situ systems. We discuss this in a later section.

Figure 2.

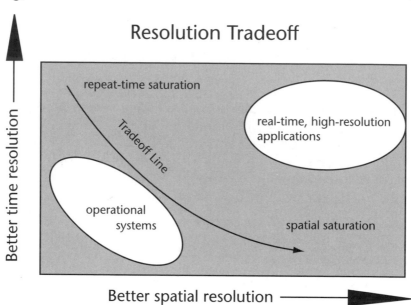

Schematic diagram showing antagonistic relationship between better time resolution and better spatial resolution. The Tradeoff Line is a fundamental consequence of the physics of the sensor technology, and the flight capabilities of the platform. In general, operational earth observing systems lie below the tradeoff line, but new technologies and processing algorithms move the tradeoff up and to the right. Some of the most useful data for humanitarian aid applications, collected and distributed in near real time at high resolution, lie above the curve. Moving operational systems into this realm is at the frontier of observational R&D.

Optimizing the Resolution Trade-off with Improved Deployments and Analysis

There are three ways of dealing with the resolution trade-off. The first is to deploy more advanced remote sensing systems, with more flexible tasking. Second, analytical processes could be designed to extract more information out of the remote sensing data already collected. Thirdly, the deployment of in situ sensor networks can supplement and in some cases replace remote sensors. Of these, the first is expensive and subject to significant technological advances, while the second and third are relatively unexplored in the context of humanitarian applications.

New procedures are being developed to extract more information from existing systems. Remote sensing data is usually analyzed by characterizing the complexities of the transmitted and reflected waves with just a few parameters. This offers the advantage of simplifying the relationship between raw data and a feature on the earth's surface. On the other hand, much of the information content contained in the wave shape, or waveform, could be lost if the parameterization is too simple. It is sometimes possible to combine many parametric observations to reduce scatter. Another technique combines the raw waveforms from many sensors before parameterization. Such "stacking" is the basis of so-called synthetic aperture systems, which often provide low-noise and high-resolution measurements of swaths of the earth's surface.

An important area of research in remote sensing revolves around the use of the full waveform. A sensing instrument receives, or transmits and receives, radiation in the form of waves. These waves are modified in somewhat predictable albeit complex ways by their interaction with the target. Rather than characterize these complexities with just a few parameters, some types of analysis use the full waveform. The full waveform inherently contains more information about the surface, but the analysis requires more computation and the underlying theory is more complex. Still, significant advances have recently been made. Under certain conditions, full waveform analysis can improve spatial resolution and surface characterization without the need for new, or more densely spaced, sensors. Still, the transition to operational capability can be problematic. Figures 3 and 4 give some specific examples of improvements obtained from full waveform analysis (figs. 3 and 4: landslide mapping with SAR Polarimetry, urban classification).

In Situ Sensors and the Resolution Trade-Off

In principle, in situ sensors do not suffer from the same resolution trade-off as remote sensing systems since they are less

Figure 3.

Radar (left) and visual (right) images of the Tsaoling mega-landslide generated by the 1999 M7.6 Chi-Chi earthquake (Taiwan). Both images are about 4 km across. The radar image, which was acquired more than 1 year after the event, has been processed to promote differentiation of bare surface (purple), forest (green), other (black), and no-data (yellow). This type of processing produces a "classification" map, and aids in the rapid and nearly automatic identification of important features. The visual data, obtained only 6 weeks after the landslide, is more difficult to classify. Figure modified from Figure 4 of Czuchlewski et al. (2004), used with permission.

expensive and can be deployed at small spatial resolution with continuous recording. They also measure the target process more directly, being more specialized and usually not suffering from transmission or proxy effects. However, regional spatial coverage can be difficult unless some minimal telecommunications infrastructure is present. Thus their deployment must be designed carefully.

A common strategy is to deploy a "backbone" of sensors connected through regional or global telecommunications infrastructure (Internet telemetry through microwave radio links or cellular networks, for example). The backbone can supply a minimum level of earth monitoring so that disastrous events can be

Figure 4.

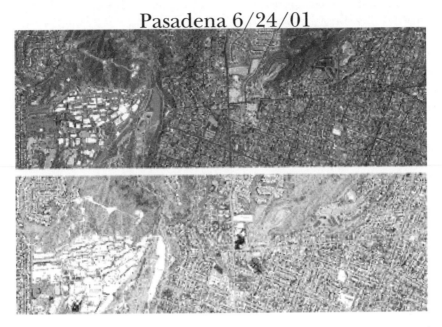

Pasadena 6/24/01

Modified from Small (2003). Classification map of Pasadena obtained from spectral mixture analysis of IKONOS imagery. This type of processing makes use of information stored in the reflected waveform, and facilitates identification of very fine-scale features such as roads. In disaster situations, variation of reflectance along identified roads may indicate damage or blockage. Figure includes materials copyright Space Imaging.

characterized quickly when they occur. Then, portable or opportunistic deployments of in situ sensors may be incorporated into the technical components of the humanitarian response to provide the higher resolution needed for decision support. This is the way modern earthquake detection networks operate, for example. In these systems, extremely sensitive seismometers are operated globally in order to quickly locate and provide preliminary characterization of damaging earthquakes in near real time. If needed, portable seismometers can then be sent to the earthquake zone to provide more detailed information on aftershocks and crustal stress changes. In areas where the earthquake risk can be accurately predicted, the backbone network can be "densified" or supplemented by additional high-resolution instrumentation. The seismographic networks in the United States, Japan, and in many developed countries operate in this mode. It is a frontier problem in both seismology and hazards reduction to identify high-risk conditions in developing and underdeveloped countries that justify densification. The same statements can be made for hazards ranging from floods and drought to landslides and volcanoes. (See Fig. 5 for global earthquake monitoring and regional monitoring comparisons in Turkey, California, and Japan.)

Ideally, remote sensing and in situ sensors complement each other. In situ sensors can be used to calibrate remote sensing observations or provide the measurements needed to interpolate

Figure 5.

GSN & FEDERATION OF DIGITAL BROADBAND SEISMIC NETWORKS (FDSN)

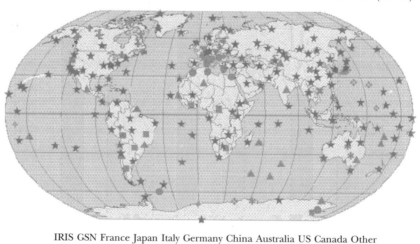

IRIS GSN France Japan Italy Germany China Australia US Canada Other

Modern seismographs of the Global Seismic Network (GSN) and the Federation of Digital Broadband Seismic Networks (FDSN). The GSN and FDSN provide global coverage of earthquake sources and allow the rapid and accurate localization and characterization of large earthquakes. In some regions, such as California, at great risk from earthquakes, the global networks are supplemented by regional or national networks located closer to potential events. These regional networks are not uniformly present in poorer regions, however, making it more difficult to make accurate forecasts of earthquake risk. Figure courtesy of the Global Seismic Network Program of the Incorporated Research Institutions for Seismology (Rhett Butler, personal communication).

between time-separated remote sensing observations. Some phenomena, such as the relatively high-frequency seismic radiation that causes damage to buildings, can only be measured in situ. In situ observation is also a necessary component of disaster response so that the human loss can be measured accurately and incorporated into loss models.

DATA MANAGEMENT AND INTEGRATION

The raw data produced by earth observation technologies must pass through a series of quality control and calibration procedures and be archived before it is released to user communities. The raw data is accompanied by metadata, which keeps track of associated information about the data itself. Both the raw data and the metadata can be useful to humanitarian aid workers.

Data from different platforms should be integrated or managed coherently according to the needs of various stakeholder communities, especially if different platforms observe the same phenomena. Additionally, humanitarian action can benefit from the integration of geophysical data with socioeconomic data. Socioeconomic variables come in many forms, of course, only a few of which can be measured by earth observation platforms. Other sources of socioeconomic data include census organizations, surveys, international NGOs, and development organizations. The quality of these data is highly variable, but there are ongoing efforts to compile globally self-consistent sets of the most important socioeconomic data. (For an example of the integration of socioeconomic data with geophysical observations, see http://www.ciesin.org.)

Data Products and Product Integration: The Community Interaction Model

Improving disaster resilience through the use of earth observation technologies requires substantial communication between humanitarian workers and the data providers. "Data products," that is, information derived from raw data through some processing or analysis, are often provided by government and commercial institutions in order to meet the needs of end users. However, the design of data products often begins with the data collecting agencies and institutions themselves. It is important to ensure that the humanitarian aid community has the base of knowledge needed to interact comprehensively with the data providers, so that higher-level outcomes can be produced. The humanitarian community and the earth observation community together should devote resources to ensuring that there is a knowledgeable level of interaction between data providers and information users. Information content will be the lingua franca of this interaction, and useful information products will be the result. The design of information products is thus mediated by the users of the information.

One way to approach this ideal state is through a "community interaction model," wherein the design and deployment of earth observation technologies and the data products produced from the observations are mediated by linkages among observational scientists, scenario modelers, and the humanitarian community that intends to use the information. A three-level hierarchy informed by humanitarian needs could thus be established. First, there should be interdisciplinary linkages among the disciplinary scientists who may use the observations to constrain their simulation and prediction models, focusing on human impacts. Next, the science needs to be linked to technology that can mitigate or alleviate human suffering. Third, these solutions need to be linked to the policy, risk management, development, and aid communities. If information products can be generated in near real time (real-time monitoring of disastrous situations) then links can be established between the earth observation and the emergency management and response communities. This needs to be done in an applications environment (with workshops, training sessions, pilot programs) and should be coupled with development and capacity-building efforts in countries where disasters are most likely to happen.

The advantages of building these interactions between the scientific and humanitarian aid communities are manifold. First, the feedback between the communities is important for modulating or giving perspective to the science and technology outcomes derived from basic research using earth observations. Second, community interactions, when continually maintained, reinforce and build the knowledge base at the high level needed to sustain the most productive use of information products. When the end user community is knowledgeable, their heightened curiosity hastens applications of new science and technology, advances technology transfer, and opens new lines of scientific inquiry. Finally, one can hope that building these communities will ultimately allow them to move beyond a simple product model of interaction to a much more beneficial mutual learning environment.

An important outcome of community interactions could be, in fact, a renewed call to integrate socioeconomic and geophysical data in meaningful ways. Without exaggeration, the need to integrate human impact data with earth observation is perhaps the signature requirement of humanitarian aid applications. Generally, the data needed are of two types: (1) background, or baseline data on population, habitats, land use, economic activity, and other socioeconomic variables; and (2) pre- and post-event "change data" or assessments of disaster impacts. Thus it is important to know, for example, how many people might be affected by drought, where the impact "hot spots" are, or where geophysical observables indicate the potential for disastrous events in the future. Some of these data (such as land use and proxies for population density) are available from the analysis of remote sensing data, for example. Others require a different data collection apparatus and appropriately designed data management systems. In many cases, the relevant data are collected by national or commercial interests that, for their own reasons, may want to keep the data proprietary.

This problem could be approached by a systematic global effort to characterize human variables and integrate them with earth monitoring. The motivation is simple: spatial data integration is a necessary component of natural hazard mitigation and preparedness, and integration in real time is a necessary component of humanitarian disaster response. Moreover, proprietary data collection and products, particularly in proprietary formats, are an impediment to the ease of acceptance and accelerated utility of data for humanitarian needs. The relationship between intellectual property and humanitarian needs for earth observation must be addressed in this context.

These discussions reflect organizational, economic, and political considerations, but there are technical issues as well. The integration of socioeconomic and geophysical data would benefit from the adoption of open interoperability standards for spatial data. Such standards provide the protocols that allow data sets collected by different organizations and under different frameworks to be managed in ways that are transparent to the end users. Standards objectify the information content of the data and make the information content, rather than the raw data, the basis for building information products and technology transfer. Community interaction in developing these standards is necessary so that the right information needs are identified. Once implemented, standards promote rather than impede the rapid development of new applications.

Information Technology Requirements for Implementing a Community Interaction Environment

Significant attention has been focused on the technology requirements needed for the efficient and rapid utilization of earth observation data. Table 1 lists a few of the elements needed to promote humanitarian use of earth observation information:

The environment for information technology is constantly changing, of course, as new innovations move to the market. Table 2 lists some of the technologies that may be relevant to humanitarian assistance in the event of a crisis or disaster. Coupling these technologies to the data streaming from earth observations systems should be a high priority.

Advances in the Natural Hazards Science and Technology Environment Relevant to Humanitarian Action

Advances in the science of natural hazards and the technologies needed for mitigation and quick response are coupled with advances in earth observation in several ways. Together, these have the potential to be beneficial to the humanitarian aid community. As usual, the difficulty lies in being able to match humanitarian needs to the best available science. Rapid transmittal of new scien-

Table 1.   Information Technology Requirements for Integrating Earth Observation with Humanitarian Action

Spatial data integration
Monitoring and near-real-time data assimilation of time series
Data QC, preliminary analysis, archiving, management for research and products
Servicing decision pathways and community interactions
      Mitigation planning
      Emergency response
      Remediation
Physical descriptions: characterizations and models
Modeling and simulation codes and results
Scenario building, description, and dissemination
Data integration from other components
Capturing feedbacks and change data
Formal and informal education
Community outreach
Products in real time

Table 2.   IT Environment for Humanitarian Assistance

Ubiquitous wireless, and self-organizing, high-bandwidth networks
Portable and handheld devices with integrated networking and communications
Web-based applications for accessing and analyzing data products
Ubiquitous computing power
Real-time sensor networks or webs
Highly accurate time and location determination with GPS and differential GPS
Digital maps and routing software
Geo-coded baseline information on natural state and socioeconomic conditions, coupled with the ability to detect change
Standards for geo-spatial data interoperability
Digital libraries

tific results to the humanitarian aid community must be carefully pipelined and targeted so that irrelevant information will not overwhelm aid organizations. Some of the most important and relevant science is oriented toward improving the skills needed to predict individual events or to forecast probable occurrences.

As theory and observation advance together, scientists and engineers are gaining improved predictive skill based on data-calibrated, physics-based models of earth processes. In some cases, new precursors that accurately predict the scope of future hazardous events have been discovered (see for example: Stark 2003; Parsons et al. 2000). As a result, scientists and engineers are placing greater emphasis on the need for scientific monitoring and data integration. The need for monitoring for humanitarian purposes could be linked to this scientific justification.

Some areas of progress in earth science and engineering that are relevant to humanitarian aid include:

1. More interdisciplinary research
2. More active design and modeling of scenarios
3. Better human impact modeling and prediction
4. Assessment of direct and indirect impacts of disasters
5. Development of early warning systems where feasible
6. Developing performance metrics for disaster mitigation measures
7. Monitoring of structures and infrastructure

One example of new predictive and integrative science, engineering, and social science is illustrated by Smyth et al. (2004). In this work, Smyth and his colleagues use a scenario earthquake near Istanbul to predict the probability of damage to a specific type of apartment building in the city. They then calculate potential damage probabilities of different structural retrofitting strategies using sophisticated computer modeling. The construction costs of these strategies are then priced realistically. When combined with earthquake probabilities derived from geophysical modeling, the cost of retrofitting can be compared against the probable benefit of preventable damage. Smyth et al. (2004) demonstrate that readily applicable assumptions lead to a positive return on the investment in mitigation, a result that could calibrate direct investment in risk reduction, another form of humanitarian intervention. In another application, similar calculations demonstrate that almost any investment in critical structures such as schools is a justified target of international investment. There are other examples (see Miletti, Quarantelli for examples).

Scientific Data Environment

Issues of scientific data management and ownership often enter into discussions of earth observation data products. The earth observation data needed by the humanitarian aid community are owned by many public and private institutions, with complex relationships among them. Proprietary data formats, security classification, improper or inadequate quality control and archiving, and unanticipated hardware or sensor limitations all raise barriers to data integration even before humanitarian needs are considered. In addition, there are significant issues related to use and storage of legacy data, for which no universal solutions exist.

Some of these issues could be addressed in a regional rather than global context. High-risk regions could be identified as areas where targeted data-gathering missions could be established before disasters occur. This should be incorporated as part of the capacity building in an economic development plan. In-country scientific and technical capacity and resources should be developed so that local stakeholders can participate in developing the information needed for future disaster response and mitigation.

DEALING WITH UNCERTAINTY: ROLE OF PREDICTIVE MODELS AND SCENARIO BUILDING

One of the most vexing issues in the management of future disasters is the treatment of uncertainty and its effect on natural hazard risk management. While the occurrence of a disaster triggers response, there are still deployment and response decisions that need to be made in the face of uncertainty, inadequate information, and even chaos. When humanitarian action is the required result, the communication of scientific uncertainty is meaningless unless the effect of uncertainty on specific courses of action is known. This is a complex topic deserving a more thorough discussion, but some general comments can be made.

It is useful to view uncertainty in terms of its propagation through the decision-making process. Scenario models and counter-factual studies are useful tools that allow decision makers to see the impact of randomized events or uncertainty on the outcomes of actions. Such modeling is becoming increasingly more realistic, as earth observations are used to calibrate the scenarios. Universities have an important role to play if humanitarian aid decisions can be incorporated into the scenarios.

Unfortunately, the disaster research and management community has a long history of scenario modeling for physical impacts but has not yet integrated its research with the knowledge possessed by the practicing humanitarian aid community. While improvements are made in the methodologies for estimating the direct and indirect losses resulting from disasters (methodologies that are beginning to incorporate more and better socioeconomic data), the role of uncertainty remains a frontier research area in decision sciences. There are a few instances where decision theory is being applied to environmental issues such as disasters, but more needs to be accomplished.

Two general areas of current research are: (1) the role of uncertainty in choosing disaster mitigation alternatives; and (2) decision making in chaotic, time-stressed circumstances. While there are many motivations for the latter, the former research area is quite subtle, working at the boundaries between motivation, choice, fear, and rationalization. Some recent progress has been made in the area of mitigating disasters and climate variability, but it is not clear that practical applications to humanitarian aid are widely studied.

Discussion: Moving beyond Natural Disasters: Technology and Humanitarian Aid:

Why should earth observation in support of humanitarian action be viewed through a disaster lens? We make no claim that humanitarian crises are triggered only by natural events. However, the technologies that are allowing us to make progress in the study of natural disasters and environmental stress and their impacts should produce information products that can help humanitarian activities in general. Perhaps discussion of the ways in which earth observation can be used for disaster mitigation and response will illuminate how these technologies can be used in other humanitarian crises.

We might approach this problem by relating disaster resilience to poverty reduction and sustainable development. In this context, disaster resilience mitigates or reduces the need for humanitarian interventions. Disaster resilience can be defined as the ability of a civil society to sustain an appropriate living standard through environmental stress or to regain sustainable growth through disaster recovery. These qualities can be themselves directly connected to the concept of sustainable development. Alternatively, the achievement of disaster resiliency can support sustainable growth strategies or can be an agent of a generalized investment in sustainable growth. In either case, an individual or collective freedom from disruption by natural disasters and environmental stress can be viewed as an individual or collective human right and thus a target for humanitarian investment. Viewed in this light, disaster resilience enhances social development, is an agent of poverty reduction, provides additional incentive for international investment, and provides a performance metric for infrastructure and social investments.

It is important to distinguish between the raw data provided by earth observation systems and the information that is needed for humanitarian applications. Most earth observation systems are constructed in response to scientific, national security, or commercial needs; the broader needs of the humanitarian community must be met by data processing applications that produce useful information. The utility of the data for humanitarian applications must depend on the information products that can be derived from the raw data. Thus the question of application of earth observation technologies to humanitarian action is really a question of whether appropriate information applications exist. Nonetheless, the types of information applications that can be written depend on the potential information content of available observations systems. Beginning the broader discussion by utilizing a natural disaster context may be a productive strategy.

Investments in the applications that provide useful humanitarian information have not been high priorities. Therefore, a significant conclusion of this paper is that these investments must be made in ways that ensure that the information is most useful. These information products may be broadly applicable to a variety of situations, or they may be specific to a particular phenomenon or application. Nevertheless, there are certain attributes that are common to many systems.

One example is the integration of land-use planning with earth observation and scientific predictive capability. An important factor in humanitarian risks is the concentration of people and cultural assets in urban areas and the agricultural hinterlands that support them. Earth observations contribute to land-use planning in three important ways:

1. By supporting a strategic planning process that addresses hazard mitigation and other environmental constraints as a component of land-use planning
2. By calibrating market and regulatory risk management strategies to real-world inputs
3. By providing support for emergency preparedness, response, and reconstruction activities, and other real-time or near-real-time decisions

Balstad Miller (2001) gives a technical overview of remote sensing technologies and their broader applications.

CONCLUSION

There is undoubted potential in linking science and technology, particularly in earth observation, to humanitarian action. Some work has been done, and more needs to be accomplished, but some lessons have been learned.

Most significantly, the inherent complexities of the problem demand an integrated multidisciplinary approach, across many types of institutions, that includes a strong component of social science. In particular, local institutional and community partnerships are crucial to the study, implementation, and success of different approaches to the research and its application. Existing local capacity in different disciplines must be integrated and reinforced to deal with complex systemic issues. While data issues are paramount to the research, implementation requires, in addition, a coordinated approach to building community.

Some conclusions relevant to humanitarian aid for disaster applications are:

1. Identify high-risk regions where disasters are probable and where they could trigger a humanitarian crisis.
2. Develop pilot programs in each of these "hot spots" that address issues of data collection, integration, and science-supported decision making.
3. Link earth observation programs to long-term investments in sustainable development.
4. Develop humanitarian mitigation strategies in addition to humanitarian response protocols.
5. Use regional workshops to promote science-supported humanitarian goals.

In the earliest days after the military conquest of Iraq, the pressure was on the Coalition to restore power, especially electrical power. As this aim proved impossible, the request was modified, the demand lowered to restoration of electrical power "at least to the level provided by Saddam Hussein immediately prior to the war." For many Iraqi householders, democracy, no matter how desirable, was not to be achieved at the expense of day-to-day minimum requirements.

Electrical power drives the turbines, lights the lamps, pumps water, and keeps the factories working, the schools open, the hospitals functioning, and homes cool or heated.

In the daily briefings held by the military in Iraq, as much emphasis was placed on energy production as on troop deployments. A major daily yardstick of the success or failure of the postwar effort was the megawatt output, which at the end of the war was 1,275 MW, a figure that was 27% of that achieved in the Saddam era. This figure rose steadily to more than 4,000 MW but still has a long way to go to reach the true requirement of 7,000 MW.

The importance of the production of sufficient energy was immediately understood by the perpetrators of the terrorist activity as they focused their destructive attention on damaging or destroying energy-producing equipment from electricity cables to pylons, from generating plants to distribution units. Each attack disrupted the daily life of the Iraqis, lowered their morale, and increased their animosity toward and frustration with the Coalition forces. The deployment of a local "power police" to guard sensitive energy sites diminished but did not eliminate sabotage.

The significance of the role of the energy sector is that more than one third of the USAID reconstruction budget in Iraq was allocated to urgent power projects.

—Larry Hollingworth

Energy Technologies for Humanitarian Purposes

Ralph B. James, Ph.D., and Helen Todosow

ENERGY IS OFTEN SAID to be the lifeblood of industrial civilizations. In fact, energy consumption is frequently used as an index of prosperity. Clearly, energy is the enabling means by which social, industrial, and culture growth can be maintained. Energy is also critical to sustaining human survival and eliminating poverty and hardship at very fundamental levels.

When man-made or natural disasters occur, basic human needs have to be met with great immediacy. Energy, whether it is for cooking, heating, providing electricity, or for medical services or transportation, becomes an essential element in enabling and meeting critical human needs during emergency and post-emergency phases of humanitarian intervention.

Available energy technologies are obviously part of the solution to these critical human needs. But the kind of energy, its source and the purpose for which it is used, how quickly and easily it can be deployed, and the good that it ultimately brings is determined by a very complex, sociopolitical and cultural context that needs to be considered. Cutting-edge advanced energy technologies do not necessarily offer the best solutions throughout the continuum of humanitarian assistance. Sometimes simple approaches and product improvements create significant differences in the human condition. However, if long-term sustainable improvement in the human condition is to be achieved and future humanitarian intervention is to be avoided, then long-term growth in human productivity is desirable. This improvement in human productivity is closely linked to available energy technologies, and these technologies need to be carefully selected and integrated in order to create ecologically and socially balanced solutions for future sustainable development.

The aim of this chapter is to provide a brief overview of energy technology issues and to identify key *high-impact* energy technologies that could provide life-sustaining and practical solutions during the emergency and post-emergency phases of humanitarian activity. The focus is on energy technologies for the domestic, commercial, and light industrial sectors. The electric utility and transportation sectors are outside the scope of this review. In addition, some evolving and advanced energy technologies will be briefly examined as technologies for potential use in humanitarian activities. Furthermore, a list of select resources is provided to aid in the identification and selection of energy technologies and to facilitate access to technical information and expertise. The reader is reminded that due to the dynamic and constantly evolving nature of energy technologies and product availability, a printed publication can only serve as a very basic guidance tool. Readers are encouraged to use the References at the end of this chapter and to consult with technology specialists in order to get current information.

Background

Fulfilling fundamental human needs is the primary objective of humanitarian activities, while improving the quality of life is the fundamental aim of sustainable development processes. Both humanitarian and sustainable development goals are closely aligned, and they need to be implemented in an environmentally sensitive manner. The cultural norms and habits of affected populations can vary greatly, as can socioeconomic conditions. Some people are used to cooking on primitive stone stoves; others are accustomed to full electrification in their homes. These factors need to be considered. The humanitarian objective must be understood as the capacity to provide basic human needs that will set people on a path to social and economic recovery in an environmentally sustainable manner.

The significance of energy as an essential element in economic growth, as well as in improving the quality of human life, is well established. However, we sometimes forget the depth and breadth of the interconnections between economic, technological, environmental, and sociopolitical aspects of energy issues. Current history is full of examples of just how disruptions in these inter-connected systems can quickly bring on acute human needs that require humanitarian intervention. Whether natural or man-made, these disruptions inevitably cause a further breakdown in the traditional energy supplies, and environmental and health consequences start to spiral out of control.

Counter to current popular thinking, civilization is not in im-mediate danger of running out of energy, including oil, at any time in the near future.[1] Currently, developing countries having 80% of the world's population but are consuming only 30% of the world's commercial energy.[2] Logically, most growth in global demand for energy in the decades ahead will be in developing countries—provided they continue to develop.

Actually, some studies in the early 1990s predicted that in twenty years the developing countries would consume as much energy as the industrialized countries.[3] This projected growth was based on the assumption that developing nations would not have access to efficient and environmentally smart energy technolo-gies, but that they would have sustained access to their conven-tional energy sources. These assumptions have not materialized, in part due to the growing disparity between the technologically "endowed" industrialized nations and the worsening conditions in some of the developing nations caused by a depletion of indig-enous energy supply, rising energy costs, and imbalanced access to energy resources. This ever-widening gap in the standard of living is occurring between industrialized and developing coun-tries, but the change is most notable in the developing countries, where poor populations are preoccupied with finding enough en-ergy for cooking, obtaining water, and other activities essential to survival.

When human resources are continually spent at a dispropor-tionate rate to other available resources, or when access to those other resources is simply not available because of natural or man-made events, a human tragedy begins to unfold. During moments of critical intervention, key steps can be taken to help people make smart choices and to take the right path forward using tech-nologies that make the most of indigenous resources and human capacity, and that are environmentally and economically sound and sustainable. This means that humanitarian organizations need to know which energy technologies are best to use and how to deploy them within the geographic and cultural context. This is not an easy job. This publication is intended to help with this process by defining some of the key issues, technologies, and resources that are available to field workers.

The United Nations 2002 World Summit on Sustainable Development addressed the Earth's most pressing problems—poverty, hunger, disease, environmental degradation, social injustice, economic inequities—and published its "Plan of Implementation."[4] Energy was a fundamental interconnecting issue that was discussed during the summit, and some of those energy issues are summarized in a brief report, "Energy Issues at the United Nations 2002 World Summit on Sustainable Development."[5] Some of the basic energy "truths" that are highlighted in this brief report, include:

- Energy is the means to every sustainable development.
- Access to energy is essential for poverty alleviation.
- The world needs safe, accessible, and economically viable energy—in all its forms—to meet environmental goals and challenges.

However, a guide on how humanitarian field workers can translate and apply specific recommendations in concrete and relevant terms was not within the scope of the summit publications.

To meet these goals, energy technology choices may be categorized into two main sectors: the fossil and conventional fuel sector, and the renewable energy sector. Both energy choices can be used effectively to reduce or eliminate a wide range of adverse human conditions that are found in many parts of world.

An Overview of Fossil Energy Issues and Technologies

Fossil fuels are still the world's most abundant energy source, and until other energy sources and technologies become more readily available and competitively priced, their use will continue to dominate. The worldwide use of coal, oil, and natural gas and their derivative products contributes to a number of environmental and pollution issues, such as acid rain, air pollution, and the accumulation of greenhouse gases, which are considered to be the prime contributors to global warming. In fact, the direct combustion of fossil fuels for transportation, heating, and industrial processes accounts for more than half of all greenhouse gas emissions worldwide.[6] Energy-related local and regional pollution is a fact of life throughout many areas of the globe and is a prime contributor to significant ecological and human health problems.

Acid rain is generally brought on by the emission of sulfur oxides and nitrogen oxides during the combustion of fossil fuels. The level of adverse effects brought on by acid rain can vary tremendously due to the variability in local and regional vegetation, soil, and weather conditions. Northeast Asia is a prime example of this.[7] The rapid growth of fossil fuel use is having a direct impact on local and regional economics, and it is adversely affecting health and the environment with a global ripple effect.[8]

Fossil fuel availability is not distributed evenly worldwide, which causes economic and sociopolitical imbalances and stresses. Furthermore, the delivery and distribution of fossil energy is a complex matter relying on energy-consuming transportation systems and complex infrastructures. The simplest end-use fossil fuel forms, such as kerosene, propane, cooking and heating oils, gasoline, and so on, are products that require a complex interplay of resources and technologies to produce. Fossil fuel's ultimate conversion to usable forms for widespread residential, commercial, and heavy industrial applications, such as electricity, is an even more involved and complex matter, relying heavily on vulnerable infrastructures and sophisticated industrial and human infrastructure support systems.

Fossil Fuel Technologies

For domestic, residential, and light commercial use, state-of-the-art fossil fuel-based technologies should be modular, transportable, efficient, and safe, and whenever possible offer dual functionality, such as combined heat and power for larger scale applications.

The most common combustion devices in the world are household stoves. They are prevalent throughout developing countries. Household stoves typically use fossil fuels, including LPG (liquid petroleum gas), kerosene, and methanol, and biofuels such as wood and plant products. Their preponderance and availability make them easy acquisitions for humanitarian purposes. But many of these stoves emit noxious pollutants, such as CO_2, CO_3, CH_4, N_2O, NO_2, and SO_2, as well as high levels of total suspended particulates (TSP), and they are the primary contributors to serious health problems and indoor and regional air pollution and environmental problems.[9] In crowded refugee conditions and within temporary shelter structures, these health and safety issues are exacerbated, and they are a prime concern.

In fact, cooking smoke has been recognized as a silent killer in developing countries. Estimates are that more than 2 million people die each year from inhalation of smoke from cooking fires. Deaths from indoor air pollution exceed deaths from tuberculosis, malaria, or AIDS, according to the World Health Organization. Indoor air quality health issues are exacerbated in colder climates, where people spend longer hours within homes or shelters. There are also additional serious safety issues, such as burns and household fires; these health and safety problems disproportionately afflict women and children. Though humanitarian activities are often viewed as temporary activities, due to circumstances they often continue for prolonged periods of time, and this offers the opportunity to introduce products that can significantly improve living conditions.

Traditional stoves/fuels are only 10–20% efficient and produce high levels of smoke, while kerosene, LPG, and electric stoves can improve cooking efficiency to 40–60%, greatly reduce smoke, and cut down on the time and human energy required to perform necessary household functions and sustain operations.

For humanitarian activities, consideration should be given to using high-quality fuel stoves that use liquid and gaseous fuels, and stoves that offer backup fuel supply options, such as wood/gas stoves.[10] Such new stoves have very high efficiencies compared to traditional stoves that are limited to biomass or wood. In general, the ranking follows what has been called the "energy ladder" from lower to higher quality fuels, with emissions decreasing and efficiencies increasing: dung—crop residue—wood—kerosene—gas—electricity. Methanol, manufactured either from natural gas or biomass, is another widely available alcohol fuel suitable for household and indoor use.[11]

Figure 1.   Emissions Along the Energy Ladder

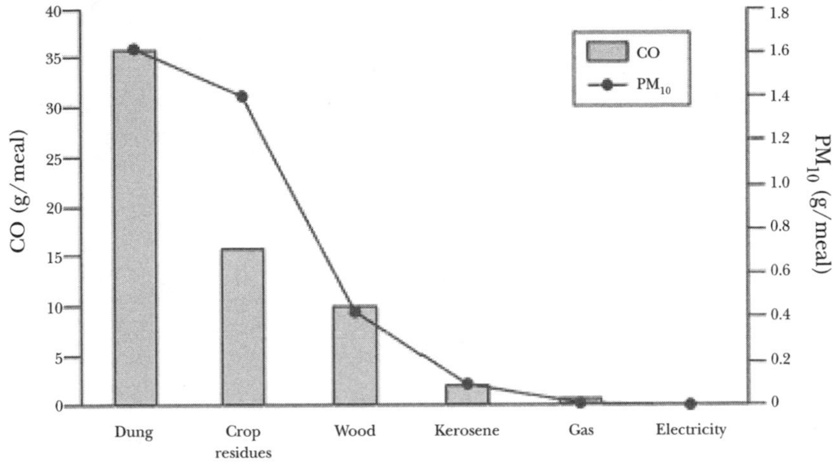

Source: ITDG "Smoke: the Killer in the Kitchen." World Health Organization: http://www.who.int/indoorair/en/)[12]

For larger scale electric power supply in support of humanitarian activities, micro turbines can play a significant role. These mini power plants can run on high-energy fossil fuel types, such as natural gas, methane gas, diesel, and propane. They offer environmentally friendly power with very low levels of nitrogen oxide, carbon monoxide, and sulfur dioxide emissions—thereby reducing greenhouse gases. Typically they are small, natural gas turbine engines from 25 to 500 KW, approximately the size of a large refrigerator. Micro turbines offer a number of potential advantages over other technologies for small-scale distributed power generation. These advantages include having a small number of moving parts, which make them easier and less costly to maintain; being compact and lightweight; and having greater efficiency, lower emissions, and lower electricity costs. In addition, they can utilize waste fuels and offer waste heat recovery, known as cogeneration and combined heat and power (CHP), which can be used for heating domestic space and hot water, and running air conditioner (AC) chillers, in addition to supplying electricity. They are superefficient, achieving efficiencies of greater than 80% (when CHP is considered), compared to an average efficiency of 30% to 40% for coal-powered systems.

Micro turbines are currently becoming more popular in industrialized countries, where they are being used for on-site power generation, as backup power supplies, and as a flexible power option for connectivity to larger electrical grids. Micro turbines are a very effective power supply option for humanitarian needs. Supplying the fuel for micro turbines presents a transportation challenge in some circumstances and locations, but because they can run on various types of fuel, this should not be a major impediment. Safe storage of fuel reserves may also present challenges. Micro turbines are available today, although further improvements, choices among commercial models, and cost reductions are expected over the next few years.

An Overview of Renewable Energy Issues and Technologies

Renewable energy technologies have advanced dramatically during the past twenty-five years, and they offer some of the most exciting energy solutions for emerging economies and developing countries. They offer potential solutions for both industrial societies and for humanitarian functions. Renewable sources of energy should be considered one of the key elements in the overall strategy for delivering humanitarian assistance and laying the foundation for sustainable development.

Renewable energy is defined as solar, wind, geothermal, hydropower, and biomass/biofuels. It also includes advanced and more esoteric energy options such as nuclear, hydrogen energy, and ocean/tidal energy. From the humanitarian perspective, the focus is and will continue to be on the more conventional types of renewable energy products, such as solar, wind, biomass, and micro hydropower.

Effective implementation of renewable energy technologies occurs when the technologies are matched to abundant *indigenous* energy recourses, like solar, wind, or biomass. Renewables can also be effectively used in conjunction with fossil energy products. However, the full range and potential of many renewables is still underutilized in the developing world and in humanitarian situations, and sometimes these resources are overexploited, as is the case with biomass, because they represent a cheap and readily available fuel source. Biomass exploitation can lead to deforestation and bring on additional environmental and human stress, depending on the magnitude of the humanitarian situation.

Deployment of renewables such as solar and wind is becoming more common as the technologies mature and become more readily available. Products ranging from simple solar water heaters, solar cookers, and water purification systems to more sophisticated solar photovoltaic and wind systems for electric power production are now available. By using simple products that use indigenous resources in efficient and effective ways, humanitarian workers can make notable improvements in people's lives, lower the demand for transport of fuels, and improve health and ecology in the immediate community.

Products and technologies that are based on renewable energy sources are often highly suitable for distributed energy delivery and do not require sophisticated infrastructures and complex systems for implementation. In addition, interconnectivity and mini-grid options are available at more advanced levels for some renewable energy technologies, such as for solar and wind systems. Furthermore, their selection for initial emergency phases of operations can be scaled up to address post-emergency energy needs.

Biofuels

More than a third of the world's population (some 2.4 billion people) burn biomass for cooking and heating purposes. In the context of fuel, biomass is defined as wood, crop residues, charcoal, and dung. The smoke from burning biomass fuels causes severe health problems. But as a result of its low cost and availability, biomass is likely to remain the main cooking fuel in many rural areas of developing countries for years to come, and it is often readily available to support initial humanitarian activities.

Because energy supply tends to be one of the most serious human and environmental issues associated with refugee camps, due to the immediate need to construct shelter and supply fuel for cooking, heating, and refrigeration, the over-harvesting of local biofuels such as wood and other plant matter can occur very quickly and contribute to both immediate and long-term hardships.

Using energy efficient cooking and heating systems is a necessary first step to delivering immediate aid and sustaining human comfort for longer periods. Simple solutions such as improved and energy efficient biomass stoves may be the most practical option for cutting smoke exposure, reducing fuel waste, cutting fuel collection burdens, and controlling exploitation of local ecosystems. However, local customs for fuel use and cooking methods may need to be addressed, because the use of biomass fuels is deeply ingrained in cultural habits throughout the developing world, and some products may be more readily received than others.

Traditions and values generally influence household energy use patterns. Open fires and mud stoves are used not only for cooking and heating purposes, but they represent a certain lifestyle choice. The aroma of the smoke enhances food flavors and serves as a natural repellent against insects and animals. The introduction of any new technology, including modified stove designs, may meet with strong resistance from the communities, even under emergency conditions, because of the deviation from established customs.

For a useful guide and summary of domestic and low volume residential biofuel use issues, the reader should consult "Refugee Operations and Environmental Management: Selected Lessons Learned,"[13] which is a sourcebook compiled in response to "Towards Sustainable Environmental Management Practices in Refugee Affected Areas" (TSEMPRAA).[14] The sourcebook also includes specific low-tech cooking and heating solutions that can be used in a variety of humanitarian and refugee situations, and it addresses cultural habits, training, and other issues that may have an impact on implementation, acceptance, and use.

Besides biomass fuels such as wood, charcoal, and dung, liquid biofuels can be used for various household uses. Biofuels are fuels derived from biomass and processed to produce a combustible liquid fuel. There are two main categories: alcohol fuels, including ethanol and methanol, and vegetable oils, which are derived from plant seeds such as sunflower, sesame, linseed, and oilseed rape. Ethanol is the most widely used liquid biofuel. Most commercial production of ethanol is from sugarcane or sugar beet. Methanol is produced by a chemical-conversion process. It can be produced from any biomass with a moisture content of less than 60%, such as wood, forest, and agricultural residues.

Both of these biofuels are significantly cleaner compared to other biomass fuels, although they can be costlier alternatives. They are currently available and used as transportation fuel in either a blended or straight form. Therefore, depending on where humanitarian intervention is required, these products may be cost-effective and readily available for basic household/residential, commercial, and transportation purposes.

A further method for extracting energy from biomass involves the production of vegetable oils. The process of oil extraction is very similar to the process used to extract edible oil from plants. There are many crops grown in rural areas of the developing world, many of which are suitable for vegetable oil production. Examples include coconut, cottonseed, groundnut, palm, rapeseed, and soybean. These oils can also be used as fuel for heating and cooking. Biodiesel is a fuel with similar properties to kerosene but is derived from vegetable oil or animal fat. Unlike fossil fuels, biodiesel, in principal, can be produced and consumed indefinitely. However, biofuels and biodiesel may compete with agricultural resources when considered for long-term use.

Biofuels, such as ethanol, methanol, and biodiesel, also tend to be cleaner alternatives compared to biomass fuel forms such as wood, dung, and charcoal. Because biofuels are processed fuels, the supply, transport, and storage of these fuels need consideration. Local production facilities may not be available or feasible in emergency situations.

Despite the prevalence of stoves for cooking, heating, and small-scale commercial use, improved stove R&D is still desirable. Innovative biomass stoves are on the market, such as the Ecostove, which was developed by the NGO Prolena.[15] This innovative woodstove is insulated, and hot emissions (smoke) are vented through a chimney. The stove is sealed, preventing nearly all indoor air pollution, and reduces consumption of wood fuel by 50%. However, because microenvironmental needs are often as complex as the broader environmental concerns, there is still a strong need for more selections in improved stove designs to meet the requirements of a wide and diverse range of cultural and regional needs and for quick deployment in support of humanitarian activities.

Solar

Solar photovoltaic (PV) and solar thermal energy technologies are extremely useful in a wide range of humanitarian settings. The growth of the solar photovoltaic industry is bringing more choices to the marketplace, with more products suitable for a wide range of uses and conditions.

Solar Photovoltaic Technology. Photovoltaic systems provide high-quality electric lighting, refrigeration, and other essentials such as water pumping and treatment, medical systems and services, communications, and security systems. PV systems are nonpolluting, and they improve indoor and outdoor air quality. For example, replacement of a kerosene lamp with a 40-watt solar module can eliminate up to 106 kilograms of carbon emissions a year, which can have a very real and immediate impact in crowded refugee conditions.

PV technology is gaining popularity as a mainstream form of generating electricity. Photovoltaic modules provide an independent, reliable electrical power source at the point of use, making them particularly suited to remote locations. PV systems are technically viable and, with the recent reduction in production costs and increase in conversion efficiencies, can be economically feasible for applications in many parts of the world.

PV solar cells can be interconnected in series and in parallel to achieve the desired operating voltage and current. Solar PV systems can be used in conjunction with other energy technologies to provide an integrated, scalable system for remote power generation. These systems are referred to as hybrid systems.[16] Common configurations of hybrid systems could include a solar PV array, wind generator, and a diesel generator, which could allow energy generation in many weather conditions. Such hybrid systems require planning but may be highly suitable for longer term humanitarian operations.

There are also simple PV products such as solar lanterns, originally created for the outdoor leisure market in western countries, that can provide very effective lighting without the use of highly combustible fuels, such as kerosene.

Solar Thermal Energy. Solar thermal energy is a low-tech option that can be used for space heating and cooling, cooking, water treatment, and crop/food drying. It offers highly effective solutions for meeting many basic human needs.

Solar cooking is a technology that has received a lot of attention and increased use in recent years in developing countries. The basic design is a box with a glass cover. The box is lined with insulation and has a reflective surface that concentrates the heat onto the pot. The pot can be painted black to help with heat absorption. The solar radiation raises the temperature sufficiently to boil the contents in the pot. Cooking time is often a lot slower than conventional cooking stoves, but there is no fuel cost—and this cooking method does not require constant attention.[17] The primary obstacle to its use is the need to limit food choices to those suitable for "one-pot" meals and to time meals to coincide with available sunlight. Both can be significant cultural issues.[18]

Space heating can be energy intensive, but building design and orientation can significantly reduce heating energy requirements. These considerations can be accommodated in some refugee settings.

There are many other uses for solar thermal technology. These include refrigeration, air-conditioning, solar stills, purification of freshwater, and desalination of salt water. JDA's novel, low-tech passive solar disinfection system "Bottles in the Sun" is worth noting.[19] JDA's Solar Water Disinfection in Central Asia (SODIS) pilot project has been adopted by other developing countries such as Thailand, Indonesia, South America, and China, and it attests to the practicality and effectiveness of low-cost creative "energy" solutions for addressing critical humanitarian needs. More information on solar thermal technologies is available in ITDG Technical Briefs, as well as from other sources noted in References at the end of this chapter.

Wind

The use of wind energy for local power production to support a range of humanitarian functions holds much promise. Unfortunately, the general availability and reliability of wind speed data is extremely poor in many regions of the world. Large areas of the world appear to have mean annual wind speeds below 3 m/s, and hence they are currently unsuitable for wind power systems; other large areas have wind speeds in the intermediate range of 3–4.5m/s where wind power may or may not be an attractive option. In regions where wind speeds exceed 4.5 m/s, wind power would be feasible. Wind speed data can be obtained from wind maps or from meteorology services.[20]

New wind technologies promise to be more efficient, robust, and compact than earlier designs. Because wind technologies are maturing and becoming more popular and cost-effective, more selections in this renewable energy category are expected in the future. Wind technologies can be used for simple stand-alone functions such as pumping water or as more complex mini-power and hybrid power systems for electric power production with battery storage options. Space and safety requirements need to be addressed. These installations are scalable and can be connected to electrical grids for distributed energy.

Wind water-pumping systems have been in use for a fairly long time in developing countries and for agricultural use in industrialized countries. More recent product developments and improvements in valves, rotors, and other components are making wind pumps more efficient, and more products are available for household or small community water requirements.

Wind pumps can also be used with a generator as part of a hybrid system to provide electricity for an electric pump, and they can be sited a distance from the actual pump. These hybrid systems tend to be more expensive but offer the advantage that the electricity can be used for other applications when not pumping and also that the electricity can be stored in batteries for use when the wind speeds are insufficient for direct electricity supply.[21]

Small-scale wind power generating systems using turbines are suitable for residential, commercial, and even light industrial applications and are considered to be a viable and sustainable renewable energy option. They already have a track record of success in support of some humanitarian efforts.[22] However, a number of notable technical challenges remain that directly affect small wind turbines, including reliability and performance issues, such as furling and yaw behavior, thrust measurements, and blade and tower loads.[23] Design reliability and performance are often not known until after several years of operation in the field.[24] New products offering integrated power production and storage systems are becoming available and worth watching due to their innovative and more robust designs.[25]

Hydropower

Hydropower is generally used for larger scale electric power production, requiring major infrastructure support, power conversion, transmission technologies, and major construction. Traditional large-scale hydropower is also associated with adverse environmental impacts, particularly with respect to fish passage and survival, water quality in reservoirs and downstream from dams, and altered flow regimes that may degrade physical habitat for a variety of animals.[26] In addition, large-scale population displacement and other environmental and ecological disruption can occur.

However, micro, mini, and pico hydropower options can be considered for field use in humanitarian activities and for post-emergency sustainable operations. For example, such small-scale hydropower plants generally produce less than 2 MW and on occasion less than 1 kW. Challenges to implementation include a number of nontechnical and technical issues, including the understanding of geology, flow, and turbulence dynamics, even though they do not require large or complex civil constructions. Micro hydroelectric power systems can produce enough electricity for a home, farm, ranch, or village. Small-scale hydropower systems offer power production, agro-processing, and water distribution options in humanitarian settings.[27] Current micro hydro systems are available as standardized kit installations. They include generators and turbines that offer "low head" and compact designs. These systems are flexible and scalable and can operate independently or connected to an electric grid.[28]

Turbine selection is a significant component to mini hydro plant functionality. Kinetic energy turbines, also called free-flow turbines, generate electricity from the kinetic energy present in flowing water rather than the potential energy from the head.

Figure 2. Micro Hydropower Plant

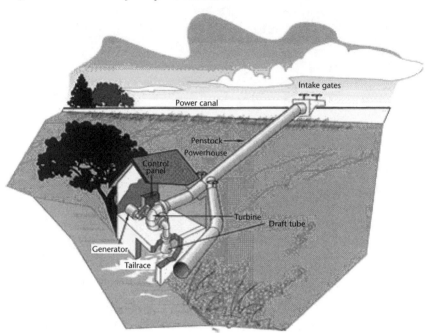

Source: U.S. DOE EERE http://www.eere.energy.gov/windandhydro/hydro_plant_types.html.

The systems can operate in rivers, man-made channels, tidal waters, or ocean currents. Kinetic systems utilize the water stream's natural pathway. They do not require the diversion of water through man-made channels, riverbeds, or pipes, although they might have applications in such conduits. Kinetic systems do not require large civil constructions, although they can use existing structures such as bridges, tailraces, and channels.[29]

ON THE HORIZON AND IN THE FUTURE

Energy technologies are one of the major focuses of research and development activities worldwide. As a result there has been major progress in increasing the commercial and technical viability of a number of technologies that could make a significant contribution to providing needed energy to serve humanitarian needs following natural or man-made disasters. These can also serve as the basis for the sustainable energy needed to allow economic growth and improvement of the human condition.

Most fossil fuel energy technology developments are focused on the power production sector, and thus they will have a greater impact on sustainable development strategies than for immediate or near-term humanitarian causes. Current research and development is focused on technologies for broad scale CO_2 management and utilization, specifically carbon capture and sequestration (for example, fixation and storage); clean, efficient, and low-cost coal utilization; low or zero emission goals; improved turbine efficiencies; and reduced water use for power production. Some of these new technologies and improvements will result in practical solutions and retrofits to existing smaller-scale fossil fuel technologies and products. Improved products, such as boilers, heaters, and others, suitable for household/residential and commercial use are expected, many of which may have relevance to humanitarian activities. This sector is very broad, with a vast array of components and subcomponents and interconnecting systems, and the reader is advised to consult References at the end of this chapter for additional details.

Hybrid systems using micro turbines in combination with batteries for storing energy or advanced fuel cells offer particular promise. Such systems could effectively meet a wide range of energy needs in remote locations or where traditional energy supplies have been disrupted or their infrastructures destroyed. Also, new developments in combined heat and power (CHP) sources will have positive ramifications for meeting energy needs in the humanitarian context.

Overall, natural gas is expected to play an increasingly important role in the development of, and transition to, an energy system compatible with sustainable development in many parts of the world. Methanol and hydrogen production from natural gas will also offer alternative power for various applications, including transportation. Methanol-based technologies are still maturing, and their use is primarily limited to the transportation sector. Hydrogen production is an exciting option for the more distant future.

Fuel Cells

A fuel cell is a device that uses hydrogen (or hydrogen-rich fuel) and oxygen to create electricity by an electrochemical process. Fuel cells have a number of advantages over conventional combustion-based technologies currently used in many power plants and passenger vehicles. Specifically, they produce much smaller quantities of greenhouse gases that contribute to global warming and virtually none of the air pollutants that create smog and cause health problems. In fact, if pure hydrogen is used as a fuel, only heat and water are emitted.[30] Fuel cells are also more efficient than combustion-based technologies, and the hydrogen used to power them can be obtained from a variety of sources, including fossil fuels, renewable sources, and nuclear energy.

Fuel cells are still far from being ready to use for widespread commercial or humanitarian purposes. There are many challenges that have to be addressed before they will be a successful, competitive energy alternative. Fuel cells are beginning to be used in vehicles and as sources for generating electricity and heat for residential and commercial uses. NASA has successfully used fuel cells to power systems in space flights.

Reducing the fuel cell system cost and size and improving the performance and durability of fuel cell systems for transportation, small stationary, and portable applications are two primary challenges.[31] Alkaline fuel cells are potentially the lowest cost fuel cell option. They are efficient (51% system), restorable (replacing the electrolyte restores output to near new), and recyclable. The reliability of alkaline fuel cells has been demonstrated convincingly over thirty years of service in space missions. New low-cost alkaline fuel cells are being developed by a number of different manufacturers. Their low cost, even in small volume production, will enable the alkaline fuel cell to be used in niche markets (for instance, peaking and/or stationary power), which is particularly suitable for developing economies and for potential humanitarian purposes.

Currently, stationary fuel cell generators for residential use are not yet on the market, although many stationary fuel cells are being researched, developed, and demonstrated around the world.[32] U.S. Department of Energy Demonstration Projects are worth watching for specific product information and developments: consult References at the end of this chapter for details.

Hydrogen

Hydrogen may indeed be a potential answer to satisfying many of our energy needs while reducing (and eventually eliminating) carbon dioxide and other greenhouse gas emissions. Several options for producing hydrogen, however, result in significant emissions of greenhouse gases. It can be produced from a wide variety of domestic resources using a number of different technologies, and it potentially offers an alternative energy source that complements other intermittent and seasonal renewable technologies. Hydrogen can be used in combustion processes and fuel cells to provide a broad range of energy services such as lighting, mobility, heating, cooling, and cooking, so it has the potential for a wide range of practical applications in the future.[33]

Today in the United States we use more than 90 billion cubic meters (3.2 trillion cubic feet) of hydrogen yearly. Most of this hydrogen is used as a chemical, rather than a fuel, in a variety of commercial applications, including methanol production, hydrogenation of fats and oils, cryogenics, welding, and so on. Currently, hydrogen's main use as a fuel is in the space program, where it fuels both the main engine of the space shuttle and the onboard fuel cells that provide the shuttle's electric power. Major technical issues have to be resolved before this technology is ready for effective land-based use, including its production, storage, and transport. But developments are worth watching.

In addition, there are safety issues. Hydrogen, like gasoline or any other fuel, has safety risks and must be handled with caution. While we have extensive experience with gasoline, handling hydrogen will be new to most of us. Therefore, developers must optimize new fuel storage and delivery systems for safe everyday use, and consumers must become familiar with hydrogen's properties and risks.[34] This factor alone makes it an improbable, though still intriguing, energy resource (other than in fuel cells) for humanitarian purposes over the next ten to twenty years. As with any new research and development area, unexpected spin-offs and benefits can occur, creating new products and technological solutions.

Ocean and Tidal Energy

Interest is growing in harnessing tidal and ocean currents and producing electrical power using this form of energy. The technology currently exists, and there are numerous sites, both in industrial and developing countries, where this technology could be deployed.[35] Offshore tidal power generators use familiar and reliable low-head hydroelectric generating equipment, conventional marine construction techniques, and standard power transmission methods.

The technology required to convert tidal energy into electricity is very similar to the technology used in traditional hydroelectric power plants. The first requirement is a dam or barrage across a tidal bay or estuary. But building dams is an expensive and very time-consuming process. Therefore, the best tidal sites are those where a bay has a narrow opening, thus reducing the length of the dam required for the operation. At points along the dam, gates and turbines are installed. When there is an adequate difference in the elevation of the water on the different sides of the barrage, the gates are opened. This "hydrostatic head" causes water to flow through the turbines, turning an electric generator to produce electricity.[36] Future technological development is likely to be based upon buoyant tethered systems and not fixed seabed approaches.

Tidal currents, although variable, are reliable and predictable, and their power can make a valuable contribution to a multifaceted electrical supply system. Tidal electricity can be used to displace electricity that would otherwise be generated by fossil fuel power plants, thus offering a clean, renewable energy option. Current technology is costly and not appropriate or ready for humanitarian purposes. However, as an alternative supplemental renewable source of energy, this technology may be promising for remote developing island and coastal communities.

CONCLUSION

As noted in the introduction of this chapter, achieving humanitarian objectives pertaining to energy supply and cost is a huge challenge, particularly in light of the potential adverse conditions when and where humanitarian assistance is needed. Since energy resources are so critical to meeting basic human needs, the selection of smart energy technology choices under emergency conditions is difficult and may involve many trade-offs. This chapter is focused on acquainting the reader with some of the key energy products and technologies that may be useful in a variety of humanitarian settings and offering the reader additional resources that can help with the decision and selection process. A few key guidelines for appropriate energy product selection are summarized below:

- Determine the indigenous energy resources available in the area or region. These can be "homegrown" resources or readily and traditionally available resources based on supply chain and/or production capabilities, in addition to renewable resources such as wind, solar, and so on. Several key resources that offer indigenous energy maps and data are included in the Resources List. Along with this information, it would be highly desirable to have access to "energy potential" maps that may guide field workers in selecting alternative renewable energy technology solutions. This can be especially useful in developing longer term energy technology strategies for specific regions and locations.
- Assess conditions and availability of energy production and supply infrastructures—if any. Repairing and/or retrofitting and/or replacing such infrastructures may be a desirable long-term goal but may not be expedient under emergency and post-emergency conditions.
- Determine key on-location or nearby contacts that could provide technical expertise. These can be government organizations, NGOs, technical institutes, universities, and commercial businesses/industries.
- Determine additional sources of expert support, such as U.S. National Laboratories, NGOs, and industrial coalitions and their members who are able to provide knowledgeable information and assistance.
- Assess human potential within the affected population that could provide knowledge and assistance.
- Assess cultural and socioeconomic backgrounds of affected population to determine past usage patterns, expectations, and custom-based skills.

- Select energy products that are low maintenance, robust, transportable, and clean and nonpolluting; use indigenous or readily available resources; and, whenever possible, serve dual functions or offer the option of scalability. Of course some products, such as stoves, do not need scalability; however dual or multifunctionality can offer backup fuel options or CHF.
- Energy options and technologies that are logistically the least challenging to obtain and supply should be considered in the overall supply strategy. However, they may not always provide the sustainable and desirable long-term solutions. For example, firewood may be locally available, but environmental depletion and human resources need to be factored into considerations of continued use of this fuel type.
- Low cost is always a consideration, but supply and logistics (transport, delivery, storage, and security) should be part of the total cost equation. The acquisition and installation of solar, wind, and micro turbine technologies may seem too expensive, but sustained access to "free" and highly efficient power production could offer the best overall economic option.
- Make product selections based on the assumption that energy needs will steadily increase once basic needs are met.

Additional technical lessons learned can be gleaned from a number of authoritative texts and web documents noted in the References at the end of this chapter and in cited references. The reader is advised that, due to the complex and constantly changing nature of energy technologies, this chapter is only meant to be *a starting point* for making informed selections of energy technologies for humanitarian purposes.

As an additional observation, further product research and development is still highly desirable in fossil and renewable energy technologies. Small enhancements and improvements, which may appear to have a limited market potential or return on investment, may have tremendous long-term advantages and value for humanitarian and sustainable development purposes. Western and other industrialized nations should consider a much broader potential marketplace for energy products. Emerging industries in developing economies offer a highly effective source for creative and practical energy products that help to meet basic human needs. Supporting this type of sustainable economic development is highly desirable.

As a final note, the authors wish to point out that products, companies, or services that are selected or listed do not reflect endorsement or validation by the authors or their affiliates. They are included as examples of potentially suitable products or services within a product category for the specified application. Recommended or UL approved products can be found in some of the sources listed in References below.

References

This list contains *select* links and references to authoritative resources that can help the reader obtain up-to-date and additional information regarding energy issues, technologies, and products, as well as identify technical experts. Selections reflect a cross section of government, NGO, and private industry sources.

CIA World Factbook: http://www.odci.gov/cia/publications/factbook/index.html. Great source of data and information on every country in the world. Topics include: geography, government, peoples, economy, communications, transportation, military, transnational, and issues. Good geophysical maps.

Development Gateway: http://www.developmentgateway.org/. Information and links to market and development activities relevant to sustainable development initiatives. Particularly useful are "Data and Statistics" and the "Gateway" to countries and "Knowledge Sharing" topics.

Direct Industry: Industry Virtual Exhibition: http://www.direct industry.com/. As the name suggests, this business-to-business website is an international directory of products, manufacturers, and industry-related information. A great deal of relevant information for the energy market.

Electric Power Research Industry: http://www.epri.com/. Though focused on the power production industry, this site offers contacts throughout the world, including Africa, Asia, and Latin America. Check out the EPRI product catalog for information on smaller-scale energy technologies.

Framework for Action on Energy, World Summit on Sustainable Development, Johannesburg, 2002: http://www.un.org/jsum mit / html / documents / summit_docs / wehab_papers / wehab_ energy.pdf. Policy paper that recaps major energy issues for sustainable development. This publication lists "UN Capacities in Energy," a useful annotated list of the UN system's involvement in energy for sustainable development initiatives.

Intermediate Technology Development Group (ITDG): http:// www.itdg.org/. ITDG "aims to demonstrate and advocate the sustainable use of technology to reduce poverty in developing countries." The website contains a wealth of information on alternative and fossil fuel issues, products, and applications for the developing world. Information is highly relevant and up-to-date for humanitarian activities. Excellent "Technical Briefs" on energy, water, sanitation, transportation, food processing, and so on. Specific product, manufacturer, and supplier information is available. ITDG publishes *Boiling Point,* a journal for those working with stoves and household energy. It deals with technical, social, financial, and environmental issues and aims to improve the quality of life for poor communities living in the developing world.

International Energy Agency (IAE): http://www.iea.org/. The International Energy Agency (IEA) was established in November 1974 in response to the oil crisis as an autonomous intergovernmental entity within the Organization for Economic Cooperation and Development (OECD) to ensure the energy security of *industrialized* nations. Good source of information on clean fossil energy and global climate change issues. Statistical energy utilization and primary energy supply data and information for non-OECD regions, including Africa, China, India, south Asia, Southeast Asia, Latin America, the Middle East, and others.

IT Power: http://www.itpower.co.uk/About.htm. A United Kingdom consultancy service in renewable energy and technologies for the UK and the developing world. Information on projects, products, and technologies.

Journey to Forever: http://journeytoforever.org/index.html. Journey to Forever is an NGO involved in environment and rural development work. Their "Projects" are particularly interesting, with links to practical and informative publications and other sources, such as information on biofuels and solar box cookers. This website has a strong focus on education.

Nautilus Institute: http://www.nautilus.org/index.html. Nautilus Institute's mission is to solve interrelated critical global problems by improving the processes and outcomes of global governance. The Nautilus Institute for Security and Sustainable Development is a policy-oriented research and consulting organization. The Nautilus Institute is active in a wide range of notable sustainable development projects, and the website has a lot of technical information in addition to discussions of policy issues. Excellent research reports and studies on energy-related matters for specific global regions and countries.

Open Directory Project: http://dmoz.org/Science/Technology/Energy/. The "Open Directory Project is the largest, most comprehensive human-edited directory of the Web. It is constructed and maintained by a vast, global community of volunteer editors." The portal offers useful links to sources on energy-related topics, including fossil and alternative energy organizations, initiatives, manufacturers, and suppliers of energy products.

Photon International: The Photovoltaic Magazine: http://www.photon-magazine.com/. Very good source of information on manufacturers, suppliers, and new products, and detailed information on worldwide demonstration projects, including those in developing countries.

Renewable Energy Policy Project: http://www.crest.org/index.html. "REPP's goal is to accelerate the use of renewable energy by providing credible information, insightful policy analysis, and innovative strategies amid changing energy markets and mounting environmental needs by researching, publishing, and disseminating information, creating policy tools, and hosting highly active, on-line, renewable energy discussion groups." Good information across all energy sectors, including solar, wind, hydro, bioenergy, geothermal, and other.

SPARKNET: http://www.sparknet.info/home.php. This is an interdisciplinary, interactive knowledge network that focuses on energy for low-income households in southern and eastern Africa. Includes a useful and easy to use database and report generation tool that provides detailed information regarding household and light commercial energy usage in select African countries.

Solar Cooking Archive: http://solarcooking.org/. Everything you may want to know about solar cooking, including manufacturers, suppliers, and recopies. A very useful directory.

Solarbuzz: http://www.solarbuzz.com. Connections to solar businesses worldwide. Contains industry, market, and business information.

Source Guide Directory: http://energy.sourceguides.com/index .shtml. A portal and guide to renewable energy business-related information. Source Guide Directories offer excellent product information and have a country index that can guide users to renewable energy suppliers, manufacturers, and assistance throughout the world.

Sphere Project, "Humanitarian Charter and Minimum Standards in Disaster Response": http://www.sphereproject.org/. The Sphere Project was launched in 1997 by a group of humanitarian NGOs and the Red Cross and Red Crescent movement. The project contains three primary elements, "a handbook, a broad process of collaboration and an expression of commitment to quality and accountability." Of primary significance is the handbook; currently the 2004 edition is available. The handbook includes a broad scope of topics related to water and sanitation, nutrition, food aid, shelter and sites, and health needs. Fuel and energy related guidance are provided only at a *very fundamental* level, for example: "Each disaster-affected household has access to communal cooking facilities or a stove and an accessible supply of fuel for cooking needs and to provide thermal comfort. Each household also has access to appropriate means of providing sustainable artificial lighting to ensure personal security . . ." No guidance or standards for selecting energy technologies is provided.

Sustainable Village: http://www.thesustainablevillage.com/about/ _about.html. The Sustainable Village website claims the world's largest selection of appropriate technology and renewable energy products, with more than ten thousand products in its downloadable database. This is an *invaluable* reference tool for all renewable energy, sustainable development, and micro-enterprise projects worldwide. In addition, the site offers a networking capability to enhance communication and provide support and training to project participants, and it is an *excellent* resource that provides solutions to global problems using renewable energy and appropriate technology. The Sustainable Village Product Catalog, http://www.thesustainablevillage.com/products/catalogs/index.html, is particularly useful for detailed technology and product information for energy, water, lighting, and other such products. Electronic sourcebooks, such as the *Micro-Hydro Sourcebook,* offer detailed technical information. http://www.thesustainablevillage.com/servlet/display/product/detail/21881.

TERI Indoor Air Quality: http://www.teriin.org/indoor/indoor.htm. Events, sources, directory of persons, organizations, and manufacturers of monitoring equipment.

Underwriters Laboratory: http://www.ul.com/dge/microturbines/index.html#contact. Lists UL Distributed Generation Equipment, technologies and components, including fuel cells, micro turbines, photovoltaics, wind turbines, hydrogen technologies, and inverters, converters, and controllers. This website provides information regarding certification, performance testing, and regulatory issues. This is an authoritative and useful source for select categories of energy products and technologies that have UL approval, it but provides limited information for some categories of products.

United Nations Department of Economic and Social Affairs, Division of Sustainable Development: http://www.un.org/esa/sustdev/partnerships/partnerships.htm. Information regarding policies and operational activities. The Energy and Transport Newsletter, published by the Energy and Transport Branch of the UN, provides insights into technology demonstrations, assessments, and market needs assessments. "With a core staff of fourteen in-house energy experts and with over 200 associated consultants, the Branch has the capacity to technically backstop projects dealing

with all aspects of this highly diverse sector around the world."
Contact: Mr. Kui-Nang Mak, Chief, Energy and Transport Branch,
United Nations, DC2–2050, New York, NY 10017, USA, Fax No.:
(1) 212–963–9883 or 9886, Tel. No.: (1) 212–963–8798, E-mail:
makk@un.org.

United Nations High Commissioner for Refugees, "Operations
Management Handbook for UNHCR's Partners," revised edition,
February 2003: http://www.unhcr.org/ . This is an emergency
preparedness handbook, offering basic guidance to field workers.
The UNHCR website also includes information on current
"Emergencies" and provides updates on news, funding, and lo-
gistics, plus regional overviews, maps, and background informa-
tion.

U.S. Department of Energy (DOE): http://www.energy.gov/en
gine/content.do. The website offers access to excellent data and
in-depth information on all energy topics, including all fossil, renewable, and emerging energy technologies. Research focus
areas provide detailed information on the following topics: energy sources, energy efficiency, environment and science, and
technology.

U.S. DOE Brookhaven National Laboratory (BNL): http://www
.bnl.gov/world/. The Department of Energy's Brookhaven National Laboratory conducts research in the physical, biomedical,
and environmental sciences, as well as in energy technologies. Of
particular interest may be the work conducted at the Energy Science and Technology Department, http://www.bnl.gov/est/
main_e.htm. Micro turbine and fossil fuel technology development and testing have been conducted at BNL. BNL is also involved in research in advanced energy concepts, including
nonproliferative nuclear fuel design and hydrogen-related initiatives.

U.S. DOE Energy Efficiency and Renewable Energy (EERE):
http://www.eere.energy.gov/ . This website offers access to the
Energy Information Portal, a gateway to hundreds of sites and
thousands of on-line documents on energy efficiency and renewable energy. All energy sectors are addressed, including buildings,

industry, power, and transportation, in all renewable energy categories. This is an excellent, extremely well organized resource.

U.S. DOE Energy Information Agency (EIA): http://eia.doe .gov/. The EIA was created by Congress in 1977 as a statistical agency of the U.S. Department of Energy. The website offers independently generated data, forecasts, and analyses and is a great source for policy making, energy market, and technical information on a broad range of energy and environmental topics. This is an excellent and authoritative source for technical data and indigenous and potential energy resource-related information.

U.S. DOE National Renewable Energy Laboratory (NREL): http://www.nrel.gov/about.html. NREL's mission is focused on developing and advancing renewable energy and energy efficiency technologies. NREL conducts renewable energy and energy efficiency R&D in the following energy programmatic areas: biomass, building technologies, distributed energy and reliability, transportation technologies, geothermal, hydrogen and fuel cells, solar, wind, and hydro, among others. This is an excellent portal to other authoritative and useful sources, including national and international organizations, technologies, news, and so on. Worth noting are NREL's National Wind Technology Center website, http://www.rsvp.nrel.gov/wind_resources.html, for international wind data, and the *Power Technologies Databook, 2003*, http://www .nrel.gov/analysis/power_databook/, a comprehensive set of data about power technologies from diverse sources.

U.S. DOE Los Alamos National Laboratory (LANL): http://www .lanl.gov/projects/cctc/ The Clean Coal Technology Compendium website contains a unique compilation of information and technologies relevant to clean coal utilization.

U.S. DOE National Energy Technology Laboratory (NETL): http://www.netl.doe.gov/.This laboratory conducts research and development in all aspects of fossil energy resources—coal, natural gas, and oil—and focuses on creating and deploying commercially viable technical solutions to energy and environmental problems. *Gas Processing* and *End Use* contains information on fuel cells, micro turbines, and other advanced fossil energy technologies. Descriptions of various demonstration projects can be a good source of specific product information, though the focus tends to be on the power and industrial sectors.

U.S. DOE Oakridge National Laboratory (ORNL): http://www.ornl.gov/. The Department of Energy Oakridge National Laboratory conducts broad-based R&D, including research on energy production, distribution, and use. ORNL's Energy Efficiency and Renewable Energy Program conducts research and development on sustainable energy resources and technologies, http://www.ornl.gov/sci/eere/.

Village Power 2000: http://www.villagepower2000.org/. "Technology to light up and link up the world." This website contains news, demonstration projects, success stories, suppliers, and links.

World Bank Group (WBG): http://www.worldbank.org/. This website contains policy, research papers, and program information for World Bank activities. The Wind Energy Resource Atlas, http://www.worldbank.org/astae/werasa/index.htm, is a useful resource for practical applications of wind technologies.

World Business Council for Sustainable Development: http://www.wbcsd.org/. A source for current information, news, events, and documents, "Energy and Climate" issues are a focus. Section on "Innovation/Technology" provides up-to-date news on emerging products and technology developments.

World Health Organization (WHO): http://www.who.int/en/. Information, publications, issues, and WHO-sponsored activities on energy, sustainable development, and health. *Methods and Tools, Publications, Related Sites,* and much more. The WHO Indoor Air website http://www.who.int/indoorair/en/ provides details on indoor air quality issues.

World Meteorological Organization: http://www.wmo.ch/index-en.html. "The UN system's authoritative voice on the state and behavior of the Earth's atmosphere, its interaction with the oceans, the climate it produces and the resulting distribution of water resources." The website offers detailed climate and meteorological data, and authoritative leads to regional and country sources of data. WMO publishes an on-line electronic publication, *World Climate News,* that contains useful indigenous energy information that can be used for energy technology deployment.

World Resources Institute: http://earthtrends.wri.org/. Earth-trends Portal offers access to worldwide climate and atmosphere databases.

Sustained and/or chronic conflict has a negative impact on access to health care by the population. This is either due to financial constraints or poor health infrastructure and health systems. Subsequently the coverage of extended program of immunization (EPI) is normally low in a conflict-affected population.

With limited EPI, there is a higher risk of an outbreak of a communicable disease. In the overcrowded camps, this is potentially disastrous. Vaccination campaigns are an attempt to increase EPI coverage and lower the risk threshold. Such campaigns require careful planning and preparation, especially adequate logistics. Some vaccines, for example, measles, are heat sensitive and not conducive to extreme temperatures. Vaccinators, equipped with carriers, filled with ice packs and susceptible vaccines work in difficult conditions. Vaccines are pulled in and out of the carriers all day, and ice packs rapidly melt and loose their efficiency. Some vaccinators are not very vigilant, and vaccines are overexposed and perish. The vaccinators do not know the difference between a good and bad vaccine and continue to inoculate regardless.

In November 2001, I was working for Médicins Sans Frontières (MSF) in the border area of Pakistan and Afghanistan. MSF was providing medical assistance and health screening to the thousands of Afghans who were attempting to flee the war and were stuck between the two countries in the no-man's-land, waiting to be processed. The MSF health center was a series of tents thrown up in the desert. The screened refugees stayed in a transit area for twenty-four hours before being trucked to the cluster of UNHCR refugee camps away from the border.

One day, with an unexpected high number of new refugees, we ran out of measles vaccine. Fortunately, the Ministry of Health (MOH) was able to assist with supplies, but there were concerns about quality, and there was no means of testing. Choices were limited; MSF had to continue screening the refugees as overcrowding in the no-man's-land was rapidly becoming an issue. The vaccines were accepted, and approximately two thousand people were screened that day and vaccinated with MOH measles vaccines. Two weeks later, the worst-case scenario occurred: three cases of measles were diagnosed in refugee camps in Pakistan. MSF investigations concluded that the three cases were not random. They were all screened on the same day and vaccinated with MOH replacement measles vaccine. Fortunately, due to a timely intervention and immediate follow-up, the situation was contained and no lives were lost. This is not always the case.

There are two immediate challenges to vaccine technology to help improve the impact of humanitarian action. The first would be to provide a means for the national health system to manage and sustain a cold chain (proper temperatures for vaccines) with appropriate technology. The second would be to help find a cheap and convenient way to consistently and quickly test the quality of vaccines during a vaccination campaign. This would contribute towards the World Health Organization's worthy campaign to eradicate certain treatable diseases such as polio.

Water, a crucial element in any environment, is especially critical in a humanitarian setting. The population in danger must have access to a safe water supply.

In 2002, I was working for MSF in South Kivu, the Democratic Republic of Congo (DRC). South Kivu was a complex and constantly changing environment, where the various armed groups converged in diverse changing alliances. Access was extremely limited, little more than a corridor, and war was still being waged in the province. Chronic insecurity and a rapidly changing front line meant that the international response was far from sufficient to cover the needs. The majority of the population did not receive assistance, and circumstances in the areas of conflict were desperate. The continuous fighting led to local conflict and the pillaging of villages, leading to a cycle of displacement, return, displacement.

Baraka was a coastal town in the south of South Kivu and was largely made up of internally displaced people (IDP) (up to 60%), rebels, and security forces. Most IDPs had not sought shelter in camps but had integrated into host communities. Conditions were cramped and up to six-to-ten people shared one room.

There was no potable water in Baraka, and the supply problem was critical. The population would take water from the cholera-contaminated lake. Interviewed families and individuals stated that they were aware of the health risks involved in drinking water from the lakes and streams but had little choice. They did not have enough means to boil the water—there was a lack of charcoal and in some cases utensils.

In October 2002, the front line shifted again and NGOs were forced to evacuate from the southern part of the province. MSF officials started to hear reports about increasing numbers of cholera cases in Baraka and surrounding villages. The health center did not have sufficient amounts of medication, the staff in the cholera treatment center had fled into the bush, and people were starting to die. With multiple front lines, semiautonomous rebel leaders, and access only by lake, the situation was very challenging. We managed to negotiate access for a boat, filled with supplies and medical staff, to take down the lake to Baraka to provide critical assistance.

The bittersweet irony of this situation: there is plenty of water in Baraka and South Kivu. However, conflict has destroyed the infrastructure, so safe water sources such as covered wells and springs have been destroyed, and people are forced to take water from the lake, which is filled with disease. The challenge to technology would be to assist in the provision of a safe water supply using appropriate methodology, which would be sustainable even during conflict.

—Nicola "Nicky" Smith

Potential Impact of Advanced Vaccine and Water Technology in Humanitarian Operations

William L. Warren, Ph.D.

IN THIS CHAPTER we will discuss how advanced technologies can be implemented and make a deep impact on humanitarian operations. Progressive vaccine and water technologies are used as the specific examples. Even though technologies are being developed to make a marked improvement in immunotherapy and freshwater sources, infrastructure hurdles still lurk in the background and often are considered an unbridgeable chasm between the introduction of a product and the consumption of it, especially for military and humanitarian missions. We take this into account and also provide concepts on how advanced technologies and/or forward-thinking government actions can be used to overcome these traditional challenges to the commercialization of cutting-edge technologies for humanitarian use.

VACCINES: INTRODUCTION

Vaccine development and testing is one of the vanguards of twenty-first century medicine because it offers the ability to provide prophylactic protection and/or cure, as opposed to simply offering temporary treatment to various diseases and pathogens. Health education for diseases such as HIV clearly won't do the trick by itself. And drug treatments—which traditionally are only moderately effective against viruses—are unlikely to ever completely beat the capricious HIV, which can sequester itself for years in the body's cells. Thus, perhaps the only chance of stopping many diseases and pathogens such as HIV is with a vaccine.

This is based on the fact that one of the most successful and widely used medical interventions is vaccines [A].[1] Currently, there are twenty-six infectious diseases that now are preventable due to vaccination [A2-5]. Humans have vanquished smallpox with vaccines; polio is likely to follow soon. Vaccines can disarm influenza and hepatitis B. But unfortunately, despite the ingenious, almost miraculous inventions of biological and genetic engineering, no successful vaccine has yet emerged for many infectious agents, including HIV, dengue, RSV, EBV, CMV, HSV, HPV, HCV, TB, malaria, and others [A1-4].[2] In addition, measles and mumps vaccines do not offer complete protection. Finally, there is a great need to devise methods to address emerging epidemics (for instance, SARS), unpredictable use of biowarfare agents (for example, anthrax), and expanding worldwide epidemics such as TB, malaria, and HIV. Therefore, health officials have a long way to go in development of efficient and rational methods to design, create, and test vaccines.

Immune System Basics

The body has two types of response to the invasion by a pathogen—the innate and the adaptive immune responses. The mechanisms of innate immunity come into play first. Innate immunity is always present and can rapidly be harnessed but does not always have the wherewithal to eliminate the infection. If this is the case, innate immune responses contain the infection, while the more powerful and specific forces of the adaptive immune responses are marshaled.

The general recognition mechanisms of innate immunity are: (1) rapid response (hours), (2) invariance, (3) limited number of specificities, and (4) constancy during the response. Patrolling scavenger cells and various other enzymes and chemicals combat a series of nonspecific, or innate, defenses. The general recognition mechanisms of adaptive immunity, on the other hand, are: (1) slow response (days to weeks), (2) variability, (3) numerous highly specific specificities and (4) improvement during the response. In this case, specific antibodies and cells are tailored for specific invading pathogens (adaptive) and always involve lymphocytes.

Generally speaking, when a pathogen, that is, a germ or virus,

Table 1.   Differences Between Innate and Adaptive Immune Responses

	Innate	Adaptive
Response time	Hours	Days
Specificity	Limited & fixed	highly diverse, improves during the course of immune response
Response to repeat infection	Identical to primary response	Much more rapid than primary response

enters our body, two things happen that involve our immune system. First, we feel sick. We may experience fever, nausea, vomiting, diarrhea, rashes, and so on. Second, the pathogen triggers an innate and/or adaptive immune response in our bodies. As the response increases in strength, the pathogens are slowly reduced in number until symptoms disappear and recovery is complete. At the cellular level, the disease-causing pathogen contains foreign proteins called *antigens* that stimulate the immune system. Once activated, the adaptive immune system can become stimulated to create other proteins called *antibodies* that attach themselves to the invading pathogen and lead to the eventual killing of it. Afterwards, "memory cells" remain in our bloodstream, ready to mount a quick, protective immune response against subsequent infections by the same pathogen. If a new infection were to occur, the memory cells would respond so quickly that the resulting immune response would kill the pathogen before symptoms showed; you are "immune" from infection.

Traditional Vaccine Therapy

Helping the immune system to combat pathogens by vaccination has been one of the great triumphs of Western medicine. Succinctly put, vaccines work because they fool the immune system into thinking it's under attack by the real culprit.

Some of the methods used to fabricate vaccines are briefly described. The six traditional approaches to making vaccines are "inactivated vaccines," "attenuated vaccines," "toxoid vaccines," "viral vectors," "DNA plasmids," and "subunit" vaccines. The "inactivation" approach, one of the better ways to accomplish this trickery, is to simply expose the body to a dead version of the virus. If the virus is intact, it will look just like the live, disease-causing version. Because it's dead, it can't replicate. But the immune system will still react, furiously churning out Y-shaped proteins known as antibodies that lock on to the virus, immobilizing it. Once the body has made these antibodies, it should remember—at least for a while—how to do it again. This is the strategy behind Jonas Salk's polio vaccine.

The "attenuation" approach also involves first isolating the specific germ causing the disease and then weakening it by aging it or altering its growth conditions. Vaccines made in this way have historically been the most successful vaccines, probably because they multiply in the body, thereby causing a large immune response. However, these live, attenuated vaccines also carry the greatest risk because they can mutate back to the virulent form at any time. For this reason, attenuated vaccines are not recommended for use in patients with weakened immune systems. Examples of attenuated vaccines are those that protect against measles, mumps, and rubella. Table 2 shows a number of vaccines and how the live attenuated or the inactivated vaccine are the most commonplace approaches to thwart disease. However, newer advanced vaccine methodologies are finding acceptance and are discussed next.

The "toxoid" approach involves injecting a toxin into the body to induce an immune response. The toxin is often treated with aluminum or is adsorbed onto aluminum salts to decrease its harmful effects. Examples of toxoids are the diphtheria and the tetanus vaccines. Vaccines made from toxoids often induce low level immune responses and, therefore, are often administered with an "adjuvant"—an agent that accelerates and increases the immune response. For example, the diphtheria and tetanus vaccines are often combined with the pertussis vaccine and administered together as a DPT immunization. The pertussis acts as an adjuvant in this vaccine. When more than one vaccine is administered together it is called a "conjugated vaccine." Toxoid vaccines often require a booster every ten years.

The "viral vector" approach takes pieces of the pathogen's

Table 2.   Various Viruses, Vaccine Approaches and Routes towards Administration

Virus	Vaccine Type	Route of Administration
polio	live, attenuated (Sabin strains)	oral
measles	live, attenuated	subcutaneous
mumps	live, attenuated	subcutaneous
rubella	live, attenuated	subcutaneous
rabies	inactivated	intramuscular
influenza	inactivated	intramuscular
yellow fever	live, attenuated	subcutaneous
varicella zoster (chicken pox)	live, attenuated	subcutaneous
rotavirus	live, attenuated	oral
hepatitis A	inactivated	intramuscular
hepatitis B	subunit	intramuscular
tick-borne encphalitis	inactivated	intramuscular

genes and cuts and pastes them into a harmless virus, such as vaccinia. The virus then infects human cells (transfection) and harnesses the cells' machinery to produce viral proteins. These proteins crop up on the surface of the cell, like beacons on a hill. The beacons alert special killer immune cells known as cytotoxic T-lymphocytes (CTLs), or killer T-cells, which are activated to destroy the infected cell.

The "DNA plasmid" approach is based on relatively new technology that harnesses DNA plasmids—rings of DNA that live in bacteria. Scientists insert pieces of the pathogen's DNA into plasmids then inject the "naked" DNA into a person. Once the plasmids enter a human cell, they migrate to the nucleus and begin expressing proteins. The proteins then migrate to the surface, as they do with viral vectors, alerting killer T-cells to destroy the infected cell.

Last, "subunit" vaccines isolate, replicate, purify, and use only that portion of a pathogen's genetic code needed to induce an immune response; the vaccine candidates are immunogenic peptides/regions of the protein, instead the whole protein. Borrowing from the strategy of the successful hepatitis B vaccine, scientists surmise that the immune system recognizes structural features on a virus, that is, pieces of its coat, in particular such as gp120 for the HIV. Using recombinant DNA technology, scientists clip pieces of the pathogen's gene that code for certain proteins. They insert the genes into bacteria, which start expressing the protein en masse. The harvested proteins are injected into the body, where it is hoped they will stimulate an immune response rendering the pathogen inactive. Subunit vaccines are also safe for patients with weakened immune systems because they cannot cause the disease. Advances in the area of antigen processing and presentation have made subunit vaccines an integral part of vaccine design strategy.

Traditional Approach Failing

Emerging infectious diseases, the development and potential destructive use of biowarfare agents, the increase in international transmission of disease due to globalization, and the worldwide prevalence and continued spread of current epidemic diseases— these are all prominent and pressing reasons to focus resources, people, and attention on improving the cycle time and quality of vaccine development.

Almost all vaccines to infectious organisms were and continue to be developed through the classical approach of generating an attenuated or inactivated pathogen as the vaccine itself, as illustrated in Table 2. This approach, however, does not take advantage of the explosion in our mechanistic understanding of immunity. Rather, it remains an empirical approach that consists of making variants of the pathogen and testing them for efficacy in nonhuman animal models.

What limits advances in the design, creation, and testing of more sophisticated vaccines? First, only a small number of vaccines can be tested in humans, as there is little tolerance of harmful side effects in healthy children due to experimental vaccines.

This limits the innovation that can be allowed in the real world of human clinical trials, with the exception of cancer vaccine trials. Second, it remains challenging to predict which epitopes are optimal for induction of immunodominant CD4 T and CD8 T responses and neutralizing B cell responses. Third, small animal testing, followed by primate trials, has been the mainstay of vaccine development. Such approaches are limited by intrinsic differences between human and nonhuman species, and ethical and cost considerations that restrict the use of nonhuman primates for trials. The outcome of these limits is a slow translation of basic knowledge to the clinics, but equally important, slow advances in our understanding of human immunity in vivo.

We need new ways to stimulate and boost the immune system. Furthermore, in light of today's twin concerns over national security and rising health care costs, we need to *develop, test, and assess* curative vaccines faster than before.

Motivation: The Need For Improved Vaccine Creating and Testing

Given the scope of the worldwide health problems caused by known and emerging infectious diseases, including the potential of novel biological warfare (BW) pathogens, it is imperative that a fresh approach be taken toward the development and rapid testing of vaccines. Vaccine technology remains largely unchanged since the days of Louis Pasteur. For instance, recent failures of attempts to develop vaccines against human immunodeficiency virus (HIV), Epstein–Barr virus (EBV), *Helicobacter pylori*, malaria, and a host of BW pathogens including anthrax and Ebola are creating a societal "pull" rather than the technology "push." Furthermore, vaccine development is hampered by staid methods to test and evaluate their efficacy. For instance, although it might hasten the process, we can't adopt the methods, now deemed unethical, of the famous physician Edward Jenner, who ushered in "vaccinology" in 1800 when he inoculated humans with cowpox in hopes of staving off smallpox. Because HIV, like many infectious diseases, eventually kills almost all its victims, no AIDS vaccine can be tested by giving people a dose of HIV.

The revolution in immunology and vaccinology (development of DNA and protein delivery systems) in the last few decades has yet to impact methods of vaccine evaluation and biological testing required for the development of truly efficacious vaccines. The science and technology development in these areas has simply been inadequate. Moreover the lack of a truly integrated teaming approach has hindered development of new systems for evaluating vaccines and optimizing host responses to vaccines.

In response to this, several government funding agencies such as the Defense Advanced Research Projects Agency and the National Institutes of Health as well as foundations are embarking on bold initiatives to improve the situation. For instance, the Global Health initiative was proposed by the Bill and Melinda Gates Foundation (BMGF) on the assumption that, with greater encouragement and funding, contemporary science and technology could remove some of the obstacles to more rapid progress against diseases that disproportionately affect the developing world.

The vaccine-related challenges developed by the scientific board are listed in Table 3 (http://www.grandchallengesgh.org).

What's on the Horizon for Vaccines

Vaccine Creation: Advanced vaccine technology is now beginning to focus on permanently curing life-threatening conditions as opposed to merely treating them. This unique approach to vaccine development will allow medical science to cure and fully repair a wide variety of human diseases and injuries that currently are susceptible only of ameliorative treatment, for example, HIV, Hepatitis B and C, type I diabetes, malaria, tumors, and wound healing to name a few. For instance, therapies will cure a variety of diseases by programming and activating the body's own disease-fighting cells, leading to improved patient care while eliminating today's costly palliative treatments such as daily insulin injections and periodic dialysis for diabetics. These advanced technologies thus serve a dual purpose: increasing quality of life for patients and reducing health care costs by eliminating recurring treatments. As one example, recurring, non-curative medical treatments such as insulin injections and dialysis account for much of our nation's skyrocketing health care costs. Despite such treatments, diabetes was the underlying cause of more than 68,000

Table 3.

GOAL: To improve childhood vaccines:
Create effective single-dose vaccines that can be used soon after birth
Prepare vaccines that do not require refrigeration
Develop needle-free delivery systems for vaccines

GOAL: To create new vaccines:
Devise reliable tests in model systems to evaluate live attenuated vaccines
Solve how to design antigens for effective, protective immunity
Learn which immunological responses provide protective immunity

GOAL: To control insects that transmit agents of disease:
Develop a genetic strategy to deplete or incapacitate a disease-transmitting insect population
Develop a chemical strategy to deplete or incapacitate a disease-transmitting insect population

GOAL: To improve nutrition to promote health:
Create a full range of optimal, bioavailable nutrients in a single staple plant species

GOAL: To improve drug treatment of infectious diseases:
Discover drugs and delivery systems that minimize the likelihood of drug resistant micro-organisms

GOAL: To cure latent and chronic infections:
Create therapies that can cure latent infections
Create immunological methods that can cure chronic infections

GOAL: To measure disease and health status accurately and economically in developing countries:
Develop technologies that permit quantitative assessment of population health status
Develop technologies that allow assessment of individuals for multiple conditions or pathogens at point-of-care

---

deaths last year in the United States and a contributing cause of death in more than another 141,000.[3] However, these statistics do not begin to tally the full economic cost to the nation in terms of lost productivity and repeated hospitalizations for those who continue to suffer. Diabetes is just one example of the multitude of conditions that the advanced vaccines will be able cure.

Vaccine Testing: While the duration and size of human clinical trials may be difficult to reduce, there are several parameters in preclinical vaccine development that may be possible to optimize further. By increasing the accuracy of the models used for preclinical vaccine testing, for example, it should be possible to increase the probability that any particular vaccine candidate will be successful in human trials. In addition, an improved model will allow the community to collect increasingly more informative data in preclinical tests—to help redesign and optimize vaccine formulations before trials.

The biologic testing of human vaccine efficacy has traditionally relied on small animal models, such as mice and rabbits, as well as more expensive and precious nonhuman primate models. While all of these models are critical for success, there are currently no sophisticated in vitro models of the human immune system. Recent efforts at DARPA, the NIH, and several foundations are well poised to meet this challenge.

Thus, the picture isn't that bleak; the field is on the cusp of an explosion of activity in the rapidly growing field of biomedicine that will revolutionize health care treatment. These technological advances can afford the opportunity to develop novel therapeutic and vaccination strategies targeting disease. For instance, much study is being done to further understand how the innate and adaptive immune responses are activated, as well as to identify and characterize those inducible genes responsible for anti-pathogen activity and the regulation of immune response.

Is Good Technology Enough to Make a Dent?

Let's assume that the academic, government, and industrial communities are able to design the much-needed vaccines. Will they make a dent in developing countries and humanitarian operations? It is safe to assume that the age-old saying "If you build it, they will come" has been shown time and time again not to be true for vaccines. For instance, one only has to track the slow uptake of hepatitis B (HB) and Haemophilis influenzae type B (HiB) vaccines, both approved in the last twenty-five years in developing countries.

A quick survey of vaccine delivery in the last few decades underscores the urgent need for advanced planning. Yellow fever vaccine, available since 1937, is used in fewer than one-third of the countries where the disease is endemic. Hepatitis B vaccine came to the front burner in 1981 but sadly reaches fewer than half of the world's children in regular schedules. The factors that account for this pertain to (1) vaccine cost, (2) delivery infrastructure and (3) burden of disease, and were the strongest predictors of whether or not HB and HiB were added to childhood immunization programs.

There are several aspects that concern delivery infrastructure that largely impact vaccine availability to developing countries and humanitarian operations. While developing, making, purchasing, and bringing vaccines to developing countries are daunting tasks, it is well known that the real challenges can start at the airport, after the vials have been unloaded for distribution. The important questions at this stage have nothing to do with vaccine technology per se but have much more to do with logistics and communications. The driving questions are: Are there sufficient refrigerators, trucks, syringes, and syringe disposal facilities for the vaccine dispensaries? Are there trained personnel to administer and handle the vaccines at the dispensaries?

The photograph in Fig. 1 demonstrates the infrastructure that is often found in developing countries to transport vaccines in a refrigerated system. This local version of refrigerated transport used a solar-powered refrigeration system mounted on a camel to keep vaccines cool on the way to remote health clinics in the east African country of Djibouti. South America and Asia are currently operating more than two hundred stationary versions of this photovoltaic refrigeration system.

Succinctly put, is the necessary infrastructure available? If the answer is no, then it will not matter whether one has the best vaccine around, or if a country has an explosive epidemic or an early one that could be stopped with a relatively small-scale immu-

Figure 1.

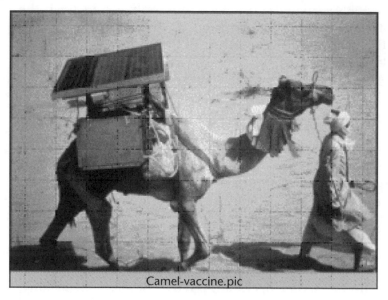

Camel-vaccine.pic

Source: Figure 1 from: http://www.worldbank.org/html/fpd/energy/subenergy/ solar/medical.htm

nization campaign. Without infrastructure, there is an unbridge-able chasm between access and delivery of the vaccine.

Besides infrastructure, one also needs to consider the issue of vaccine cost and the need for industry to make a profit. Need and demand estimates are important to the private sector when making decisions about advancing products from the laboratory to clinical trials to the marketplace. Questions such as how many of the countries that need advanced vaccines will be able to de-liver it, and how many doses will the operation need, require an-swers as industry prepares to announce new vaccine products. Perhaps the most important question: Is there sufficient manufac-turing capacity and technology to meet global demand?

Although the R&D costs to devise new vaccines can take hun-dreds of millions of dollars, it is the other costs that can become prohibitive. Building industrial capacity for vaccine production typically requires four to five years lead time, hundreds of millions of dollars, and extensive regulatory oversight. Making matters worse, the decision to build production plants must be made before the results of phase III trials are known, to avoid major delays between licensing and wide-scale availability.

For pharmaceutical giants like Merck and GlaxoSmithKline, all of the above is well-trodden ground. The hidden factor, though, is that some vaccines, including AIDS vaccines, are needed most in countries with little ability to pay. If this is the case, then it becomes important that public sector groups also receive information on demand, both to encourage investment in vaccine development and to inform plans for delivery infrastructure and financing mechanisms.

One specific program to overcome this potential problem is Project BioShield in the United States—a comprehensive effort to develop and make available modern, effective drugs and vaccines to protect against attack by biological and chemical weapons or other dangerous pathogens. Project BioShield has been implemented to accomplish the following:

- Ensure that resources are available to pay for "next-generation" medical countermeasures that normally would not have a "commercial" market. Project BioShield is designed to allow the government to buy improved vaccines or drugs for smallpox, anthrax, and botulinum toxin. Funding resources would also be available to buy countermeasures to protect against other pathogens, such as Ebola and plague.
- Strengthen government development capabilities by accelerating research and development on medical countermeasures; and
- Provide the FDA the ability to make treatments available in emergency situations—this closely controlled new authority makes available the newest treatments to patients who need them in a crisis.
- Create a permanent indefinite funding authority to spur development of medical countermeasures. This authority will enable the government to purchase vaccines and other therapies as soon as experts believe that they can be made safe and effective, ensuring that the private sector devotes efforts to developing the countermeasures.

The error of not carefully monitoring vaccine production was tragically evident in the forced recall of millions of doses of flu vaccine just prior to the 2004 flu season. A proper shield demands increased attention to all phases of production, including, in this transnational world, dependence on foreign production.

The Future Vaccine Assessment

Thus, from the above information, the outlook for advanced vaccines impacting humanitarian operations can be summed up in two words, *cautiously optimistic*. There are vast funding opportunities for advanced R&D for vaccines, which leads one to believe the vaccine scenario looks bright. However, having the efficacious product is only a part of the playing field when it comes to introducing and delivering new immunotherapies into the community, particularly the humanitarian community. Other factors such as infrastructure, cost, and patient receptiveness to the vaccine are just as important. It is possible to encourage investment in vaccine development and to inform plans for delivery infrastructure and financing mechanisms if the public sector is involved from the get-go and/or the need is high enough. Eliciting enthusiasm and dollars from the public sector can be a tortuous path, but with the right champion(s), and if the passion and need are high enough, then this too can be achieved.

WATER: INTRODUCTION

In this section we consider the importance of and necessity for freshwater and a few promising technologies that can make great leaps in humanitarian operations. The paper discusses several technological concepts that turn the water problem upside down, in a manner of speaking, by exploring avant-garde ideas to break down the energy barrier to desalinate brackish water and seawater. This section presents a discussion of the need for new water sources and a brief history of desalting research that largely finds its historical roots in the United States, followed by a succinct discussion of various promising technologies to obtain freshwater supplies that may aid humanitarian operations.

The Water Need

A growing emphasis in the foreign policy of several countries is preventative defense and environmental security as humane, cost-effective alternatives to armed conflict and intervention. Water purification and desalination are key focus areas since they not only meet future water demands globally but can also provide new opportunities in water-starved and strategic regions. The willingness of countries to meet future water demands is good news for the humanitarian community, which will be able to take advantage of the advanced technologies and resources that result from these increased international investments. The reason for the increased awareness in water is simple—water use has been increasing twice as fast as the population, and the resulting shortages have been worsened by contamination. For example, the population of the Middle East is expected to double in the next twenty years with a projected annual water deficit of more than 1.5 trillion liters by the year 2015 for just the regions of Israel, Jordan, and the Palestinian controlled areas.

The four principal science councils of the United States (National Academy of Sciences), Israel, Jordan, and Palestine have stated that the most critical of all the problems facing the Middle East is "ensuring sustainable water supplies." We believe that recent materials, nanoscience, biomimetic, and manufacturing revolutions can be applied to ionic separation water technologies such as reverse osmosis (RO) to make significant improvements in water treatment and quality that are energy-efficient and cost-effective. However, water generation and distribution has not been an area of basic research and innovation because water is believed to be a mature technology, which has stymied progress in the looming water crisis. Recent work in understanding the kinetics and energetics of desalting, the effects of charge, hydrophilicity, porosity, pore chemistry, and performing active control of the ionic separation surface is anticipated to improve the water flux by significantly decreasing energy consumption and providing sizeable improvements in our ability for in-line processing of contaminants in the water stream in a cost-effective manner.

We consider some of these recent advances and how they can impact the overall water "crisis." These advances are critically important—the cost of energy to desalt is the single most important limiting factor to creating new water supplies in many regions around the world, as 41% of the cost of operating an RO system is due to energy consumption. Thus, it is anticipated that advanced technology will make a big impact in water.

Water Desalination History

Throughout history, people have treated salty water for human consumption and agriculture. Of the earth's water, 94% is saltwater from the oceans and 6% is fresh. Of the latter, about 27% is in glaciers and 72% is underground. While this saltwater is important for transportation and fisheries, it is too salty to sustain human life or farming in general.

A major step in developing new ways to desalt seawater came in the 1940s, during World War II, when various military establishments in arid areas needed water to supply the soldiers. The Office of Saline Water (OSW) in the early 1960s and its successor organizations like the Office of Water Research and Technology (OWRT) were the lead agencies to develop the desalting industry. The United States actively funded research and development for more than thirty years. By the late 1960s, commercial units of up to 2 million gallons per day (mgd) were beginning to be built throughout the world. The technology used in these systems was mostly thermal driven. In the 1970s, membrane processes such as reverse osmosis (RO) began to emerge as alternative and more energy-efficient processes.

Originally, the distillation process was used to desalt both brackish water and seawater. This process is very expensive and extremely energy intensive, which restricted the applications for desalting to municipal purposes. By the 1980s, desalination technology was pretty much considered a commercial enterprise. And by the 1990s, the use of desalting technologies for municipal water supplies was largely becoming commonplace, especially RO processes.

The potential for large numbers of RO membrane applications is reflected in the myriad of membrane-like uses employed by nature. Membranes are used for processes such as the separation of nutrients, absorption of gases from air and water, selective protection from toxins, selection and passage of ions to transmit nerve pulses, photosynthesis, removal of waste products and toxins, and structural support throughout nature. Of particular note is the fact that all of these natural functions are highly efficient and take place under mild conditions.

Energy savings is likely to be the most significant potential advantage of using membranes over other separation techniques. For example, distillation or freezing methods of achieving separations involve a phase change, which consumes a large amount of energy. Although membrane separations can require high-pressure (energy-intensive) pumping at one or more stages, improvements in membrane surface chemistry are lowering these pressure requirements. Membrane processes can often be operated at lower temperatures, which mimic natural membrane processes. Membrane separation technologies also offer capital cost savings and flexibility. Distillation-like separation technologies are economical only in large-scale operations; membrane separation processes offer cost advantages at both large- and small-scale operations, as can be seen in under-the-counter RO systems.

Reverse Osmosis

Reverse osmosis is the industry standard to beat. As noted above, it uses membranes with extremely small pores to separate water (the solvent) from dissolved molecules (the solute). Externally applied pressure increases the chemical potential of the saltwater feed stream side of the membrane, causing the water molecules to pass through the membrane, leaving the larger salt molecules in the feed stream. To date, RO is most widely used in the desalination of ocean and brackish waters. Besides leaving the salt in the feed stream, RO also separates heavy metals, chlorinated organics, calcium, sulfates, bicarbonates, as well as other solutes from the water. Various types of membranes are used in RO: Cellulous acetate RO membranes are relatively inexpensive, reliable, and readily available. Composite membranes, though they are more expensive and are sensitive to chlorine, reject a higher percentage of minerals and tolerate a much wider pH range, which allows for efficient cleaning. Several RO polymers include: cellulose acetate, cellulose triacetate, poly(benzimidazole), poly(benzimidazolone), aromatic polyamide, and polyimide.

Energy Recovery Reverse Osmosis

Some recent advances in RO processes include energy recovery using a pressure exchanger and anti-fouling membranes. For instance, the Spectra Watermakers Clark Pump (www.spectrawatermakers.com), a new and extremely elegant device for use in reverse-osmosis desalination systems, recovers the mechanical energy from the concentrate flow and returns it directly to the feed flow as shown in Fig. 2.

The Spectra system features a unique two-stage pumping and pressurizing system. The first stage is a simple diaphragm feed pump which supplies raw water flow and pressure for prefiltration, circulation, and the driving energy for the second stage. The second stage involves the pressure amplification Clark Pump. This pump is a unique pressure intensifier that uses two opposing cylinders and pistons with a shared single rod with different diameters that in part provide the amplification. The flow diagram in Fig. 2 shows the basic process.

Water pressure from the small feed pump is used by one of the cylinders to make the other pressurize and circulate the seawater through the RO membrane. Pressurization occurs when the rod, as it is forced into a cylinder, displaces water in the closed loop circuit. The pressure instantly rises to the point where the displaced volume of water is forced out of the membrane as the fresh product. When a driving cylinder's piston touches the base, the process is instantly reversed, which means that the pressurization is almost continuous. There is no energy wasting "backstroke" like in other systems and no need for gears or crankshafts that need oil and servicing, making this a robust system.

Figure 2.

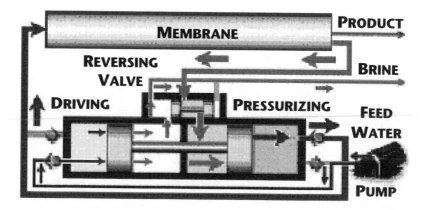

Researchers at Loughborough University have reconfigured the Spectra Watermakers system into a battery-less photovoltaic-RO system that has shown impressive performance (http://www.lboro.ac.uk/crest/Research/Project02.htm). They have been able to show a high desalination efficiency over a wide range of operating speeds. This feat enables the desalination system to be powered directly from a PV (photovoltaic) array of modest size and without need of batteries. The ability to run the system without batteries is important; the system requires no fuel and is non-polluting, and the logistics of keeping the system operational are significantly reduced. The Loughborough University study predicted the total cost of drinking water is just \$3.50 per m^3 of water, which makes the system very competitive with other small-scale seawater desalination systems. Given the large and increasing worldwide demand for drinking water, particularly in areas of medium- and low-population density, the system has enormous market potential. Furthermore, the PV-RO system could be of tremendous interest to the humanitarian community.

Even though the ER-RO system has unprecedented performance, a further advance required for this system is that of scale-up to handle larger water throughputs for larger humanitarian operations. However, the system has much promise, as it can be used to overcome logistics challenges that typically plague the distribution of water just as they can afflict vaccines, as discussed previously.

Forward Osmosis

Another advanced water purification technology that has just come to the forefront is forward osmosis (FO), which could have immense benefit to humanitarian operations as well. FO is another membrane process that uses essentially the same membranes as RO; the pore size is about five to eight angstroms. However, rather than use external pressure to push the water through the membrane and leave the bacteria, viruses, chemicals, and so forth behind, an FO system uses chemical potential as the driving force. The chemical potential is any solute (or accelerant) that has a higher chemical potential than the water to be purified. For instance, the accelerant could be dry Gatorade powder, sugared Kool-Aid, fructose, or glucose, placed inside the membrane. The sugar solution on the inside of that membrane will begin to pull water through, leaving the bacteria, viruses, and cysts outside in the swamp, as one example. Because bacteria range anywhere from 200 to 200,000 angstroms, with viruses and cysts also in there, they are too large to pass through the membrane pores. The end result is drinkable water with a slightly sweet taste that can be obtained from any non-potable source.

Another beauty behind the FO pouches is that the water source to be purified can be from a swamp, a cesspool, or even a sewage system. One adds water through the portal into the bladder surrounding the pouch. Over time, an osmotic differential is generated that pushes pure water through the membrane. The process takes no external energy. Furthermore, the process does not require iodine tablets, chlorinating tablets, nor flocculating agents.

The company developing the forward osmosis systems is Hydration Technologies, Inc. (www.hydrationinc.com). HTI has deployed the filtering system in a few hands-free products, none of which require a power source. One product is a two-liter, one-use bag that fills with water during the course of a few hours. Another is a household emergency kit that produces one liter of safe drinking water per hour, twenty-four hours a day. A third is a reusable backpack system that allows a person in the field, for example, to sip a clean drink from a bite tube while on the move. Together, the products, which have been developed in part by DARPA-funding, will likely have significant uses for the military, disaster relief, homeland defense, and outdoor recreation.

Handheld Water Purifiers

Last, another recent advanced water purification technology that has promise to impact humanitarian missions includes handheld personal-use water treatment devices such as the MSR MIOX Purifier (www.miox.com) shown in Fig. 3. The MIOX electrolytic cell uses only salt, water, and electricity to generate a liquid mixed-oxidant solution, which is injected into the water at the appropriate treatment dose. Since the primary component of the solution is hypochlorous acid (the most effective element of chlorine), a durable chlorine residual can be measured. The de-

Figure 3.

vice is ideal for travelers, campers and hikers, the military, and disaster relief.

The disinfectant will inactivate a number of common pathogens, including *E. coli, Giardia,* and *Cryptosporidium,* as well as chemical and biological warfare agents. The Purifier has passed the EPA Purifier Protocol, achieving more than ten times the level of disinfection required for normal waters. The MSR MIOX Purifier occupies only 14% of the space required by a typical portable water treatment filter and is one-quarter the weight. The compact design could be an attractive feature for humanitarian operators trying to minimize packing volume and weight who need to purify contaminated water on the spot in a rapid manner.

The Future Water Assessment

The outlook for advanced technologies to impact water supplies for humanitarian operations is very promising. As is the case for advanced vaccines, having superior technologies is only part of the equation when it comes to introducing new water-related products into the community. Logistics is still a major concern, i.e., transporting the water to where it is needed. If this is the case, there is ongoing research funded by DARPA to extract water from air. In regions of the world where sources of condensed water are scarce or nonexistent, it is highly advantageous to be able to harvest water in its noncondensed phase, for example, to extract and condense water from the air. Given the ubiquitous nature of water in the vapor phase, it is possible to establish a sustainable potable water supply at virtually any location if one can develop a technology that efficiently harvests water from air. Possession of such technology will provide a clear logistical advantage for both humanitarian workers and soldiers alike in the field by placing the water distribution capability into the hands of the person sooner. This is where up-and-coming advanced technologies can make a difference in not only water but also water logistics.

The next time you enter a large empty room or prepare to park your car in a car park, imagine that someone has laid a land mine in the room or car park. What will you do? Will you take the risk and enter or park? After all, there is only one land mine. Or will you find another room and another car park, even if the alternatives are very inconvenient for your needs? The thought of the huge blast, traumatic injuries, blood, and screams associated with an explosion will most likely deter you from entering or parking. This is the horror of land mines as they lie and wait after conflict, hidden but still lethal.

This is the situation that exists, to varying degrees, in thousands of villages and communities in many countries around the world. Their "rooms" and "car parks" are more likely backyards, fields, or grazing land needed for survival, and yet they are denied by the same terrifying thoughts of the consequences of entry.

For the last fourteen years the world has been overtly aware of this dilemma and has tried to address it in many ways. Considerable funds have been raised to train and employ people to find and remove these deadly weapons. Many real heroes are working every day to clear the land for their communities so that their families can safely return to the lives they once knew. And yet, despite this international effort, the most common way of clearing land today is still with a handheld metal detector and prodder based originally on technology that was relevant in the military clearance of minefields nearly sixty years ago.

Dedicated deminers spend hours each day wearing heavy protective equipment that is both cumbersome and hot, sometimes in temperatures up to 100°F, frequently getting up and down and frustratingly digging up distracting fragments of metal that just might have been a land mine. Their role is boring and monotonous in the extreme. Tedium sets in, and the ability to stick to safety procedures is difficult. Carelessness is possible, and yet the thought of the last accident to a deminer they knew remains with them each day. Additionally, the thought of a member of their community being injured or killed by a missed mine later motivates them in their work. Days can go by without a "real" find, and this adds to the frustration. But remember that room and that car park. You will not be happy to enter until you are totally satisfied that it is safe to do so. This probably means that someone who knows what they are doing has to check the whole room and the whole car park in order to convince you of your safety. This is the boring reality of mine clearance.

Anyone who has visited a mined area or spent just a few minutes really thinking about the consequences of living in a mined environment will readily be convinced that something should be done to improve the current methods being used, for the sake of the communities affected. If scientists can place men on the moon and achieve other incredible feats of scientific endeavor, why is it we cannot find a simple piece of plastic or metal in mud, safely and efficiently? The answer to this question is, of course, not straightforward, but technology can and should play a part in assisting the clearance of mines and unexploded ordnance in post-conflict situations.

Already a great deal of research has been undertaken and many, many ideas investigated to discover their potential. A few have survived as feasible possibilities, but few of these have evolved into developed products available to the hundreds of deminers who operate each day today. Hence, the safe return to communities, of suspected and denied land, is painfully slow.

If ever there were a need for technological solutions to a simple problem, the removal of danger, through the clearance of the debris of war, would appear to be a prime candidate.

The challenge is out there and has been for many years. Technology for mine action can, and must, help the affected communities and the deminers who risk their lives daily in the task of making their land safe.

—Noel F. Mulliner

Demining Technology

Regina E. Dugan, Ph.D.

THIS CHAPTER addresses three topics. First, it discusses the technical and cultural framework surrounding demining, "Land Mine Detection 101." Second, it considers how to measure the effectiveness of mine detectors, to demystify the scientific parameters that define detector performance, and thereby assist the nonspecialist to become a better technology consumer. Finally it reviews the history, current state of the art, and the future promise (near- and long-term) of advanced technologies important in demining and concludes with a technical addendum to help quantify and understand the technical and cultural framework of demining.

Land mines are purported to be buried in more than sixty countries and to approach numbers near 100 million. Some say that the number is exaggerated, but even if this number is a factor of two to five too high, it is staggering. In the Bosnian conflict alone, more than two million mines were laid in a country less than half the size of the state of Colorado. At present, using conventional methods, deminers remove approximately 100,000 mines per year. At this pace, it would require decades to remove those currently in the ground.

The exact number of casualties is really unknown, but thousands of people each year are victimized by land mines; in Cambodia alone one of every 236 people is an amputee. For every five thousand mines removed, one deminer is killed or maimed.

Land mines limit stabilization efforts and repatriation. They inhibit economic recovery by denying access to critical infrastructures and agriculture. They serve as the seeds of continuing terror. They are inexpensive, often costing as little as $3 to $15 dollars each, and clearing them is expensive, costing between $300 and $1,000 dollars per mine. Mines are buried in a wide variety of soils—wet, rocky, sandy, and grassy; they are plastic and metal, large and small. The United Nations estimates the cost of clearing the mined areas of the world using current methods at $33 billion.

Demining is a massive, expensive, and technically complex problem, so much so that it can freeze you in place with its sheer magnitude. But demining efforts are not frozen in place. The efforts go on, despite the small dents we currently make in the problem. Why? Because it is obviously too important a challenge to walk away from since innocent men, women, and children lose their lives and limbs every day. If you have been near this problem, it will call you to act.

I do not remember the original author or the details of the following fable, but the story seems pertinent, so I will relate it as best I can. There once was a young child who was walking along the beach after a storm. The storm had washed many, many starfish up on the beach. The child was tossing starfish, one by one, back into the ocean. When an adult observed this, he pitied the child. He saw the child's efforts as futile. So, he approached the child to tell him that his efforts would not matter, that there were hundreds of starfish on the beach. The child, somewhat astonished, replied as he tossed yet another starfish to the sea, "It matters to this one . . ." and kept on. Demining is in this mode today; we toss starfish into the ocean one at a time.

Many tasks must be completed effectively in demining. By most accounts, detection is the single most difficult task and the single most important to master. Our inability to locate mines quickly and accurately, without large numbers of false alarms, determines the painstakingly slow pace of present operations. Land mine detection is also important in military countermine operations, and this has led to many well-funded research and development efforts within the U.S. Department of Defense. As a result of these efforts, much progress has been made on the detection problem, and even recently, new technologies have emerged that promise great improvements.

LAND MINE DETECTION TECHNOLOGY

The problem of land mines is not new, and although the best systems available in 2000 were technically superior to the old

1940s-style metal detector, they were only barely so. Indeed, the only land mine detection equipment consistently used in the field still consists of a metal detector (not much different from one you might use to find coins on the beach) and a sharp pointy stick.

Even newer prototype systems suffer many false alarms. Take, for example, ground-penetrating radar. Ground-penetrating radar is a system designed to penetrate the ground and find anomalous objects, like radar systems that are pointed at the sky to detect aircraft. But the ground is very irregular, and the difference between a mine and the surrounding material is small. This causes so many false alarms that the utility of such systems is limited.

This brings up the important—indeed critical—topic of false alarms. In any sensor system—and land mine detectors are just one type of sensor system—at least two performance parameters are important. The first is the probability of detection. This determines the ability of a sensor to find land mines. If there are ten mines on a road and the sensor is able to find five of them, it has a probability of detection of 50%. The second is the probability of false alarm. This determines the number of times the sensor indicates that there is a mine when there is not. If on the same road with ten mines the sensor takes one hundred readings and indicates twenty mines where there are none, the probability of false alarm is 20%. It's important to recognize that almost any system can serve as a sensor. One could send a monkey along the road and have him randomly place flags. If he placed *lots* of flags, some of them would be over land mines. But this would be a lousy sensor. The number of false alarms would be enormous.

Ultimately, the relationship between the probability of detection and the probability of false alarm determines the quality of a sensor. High probability of detection and low false alarm rate is good; low probability of detection and high false alarm rate is bad. Scientists compare the performance of sensors to the performance of random statistical detections (like the monkey detections) to determine the quality of a sensor. We use what is called a receiver-operating characteristic (ROC) curve as a tool to evaluate and compare sensor systems. That's the take-home message of Land Mine Detection 101—one must consider detection and false alarms together in order to assess the performance of a sensor system. For more detail see the technical addendum.

Importantly, conventional land mine detection techniques do not detect what is unique to the mine. For example, a metal detector locates small pieces of metal. Some of these pieces of metal are associated with the firing pin of a land mine, but most are not. The probability of detection for a metal detector is moderate—this is what makes demining so risky. And the probability of false alarm is extremely high—this is what makes demining so slow. New technological capabilities and breakthroughs are what will fundamentally change this mix. Absent armies of deminers and droves of money, it is likely the only hope we have of transforming demining.

Today, unfortunately, metal detectors still reign supreme. As stated previously, their probability of detection is moderate and the false alarm rate is high, but there is a wealth of experience with them and they are affordable and easily obtained. Increasingly, canines are being used in various regions of the world in addition to metal detectors. Canines are the only system available today, for operation in the field, where the detection is based on what is unique to the mine—the explosive material itself. As a result, canines have good probability of detection and very low false alarms. And because they need not search every square inch of soil, their coverage rate is dramatically increased over metal detectors. Canines should be used more.

Many years ago, the South Africans designed vehicles that could withstand the force of a mine explosion. These are large, heavy vehicles with V-shaped hulls that deflect the mine blast. A group of three vehicles, fitted with large steel wheels, is capable of destroying thousands of antipersonnel mines in a single day without repair. The author herself has ridden through a minefield in the bush of Mozambique in one of these vehicles. Our vehicle withstood more than three hundred mine blasts in less than one hour, and while my ears were ringing, my limbs were intact.

An aggressive effort to use more canines and mechanical demining methods (such as mine-protected vehicles outfitted with steel wheels) would dramatically increase the safety and pace of demining activities. These technologies are available today.

When I was a program manager at the Defense Research Advanced Projects Agency (DARPA), we started the Dog's Nose Program. Motivated by the extraordinary capabilities of canines, we set out to develop sensors capable of detecting the explosive materials in mines. These systems, unlike canines, would have calibration buttons, would not fatigue, and would not suffer the corrupting influences of a rubber ball reward. Importantly, we found promise in two techniques: quadrupole resonance and "sniffers."

Quadrupole resonance is a radio frequency technique that works just like magnetic resonance imaging (MRI), only without the large external magnet typical of these fixed-base medical diagnostic systems. Advances in electronics and signal processing, a focused commitment of resources, and extensive field testing have made portable, one-sided quadrupole resonance sensors one of the most promising technological advancements. Because quadrupole resonance uniquely detects the explosive material itself, it has the ability to change radically the traditional relationship between the probability of detection and the false alarm rate in land mine detection operations. On its current development path, quadrupole resonance could be available within three years and may dramatically improve the efficacy of land mine detection. This is particularly true for humanitarian demining where the time constraints are less stringent than they are in military mine detection scenarios.

At least one type of electronic sniffer has also shown great promise; it was recently tested in Bosnia with very good results. The system was used in conjunction with a mine-protected vehicle. Air samples from wide areas were drawn into the vehicle and analyzed using a compact, relatively inexpensive sensor in lieu of a dog's nose—an electronic nose instead of a biological one. In this way, deminers could operate from the safety of a mine-protected vehicle and rapidly detect mines with high reliability. This same technique can be used to assess rapidly where mines are not and has the potential to increase ten- to one hundred-fold the rate at which areas are released as unmined. Area reduction is almost as important as demining activities in known mined areas.

We can be cautiously optimistic about other nascent technologies, such as improved metal detectors, new advanced ground-penetrating radar systems, and certain acoustic techniques. When one thinks about these new techniques it is important to remember that the challenge will be to control false alarms, since these systems have detection schemes that are most often reliant on small differences in properties related to the presence of a mine and do not focus on what is unique about the threat itself. Remember Land Mine Detection 101: ask about the false alarm rate. And think about "monkey detectors."

What might be possible in the longer term? One can envision a time wherein inexpensive small robot systems automatically search defined areas using combination sensor systems. The precursors of such automated systems exist in new home vacuum cleaners and pool cleaning systems. These systems would operate nearly twenty-four hours a day, seven days a week. They would have wireless communication to a central location that provides a detailed map of the mine locations.

The landscape of technologies that might be applied to mine detection changes over time. These changes have not, unfortunately, improved demining activities as rapidly as we may have liked. Nevertheless, each year there are at least a few new technologies noted. Most of these technologies, when brought to the field, have failed to meet expectations. It is no wonder that humanitarian workers have found it confusing and frustrating.

It is nearly impossible to be a technical expert in all the technologies that might pertain to mine detection. They vary from radar systems to chemical detection systems to genetically engineered microbes. The language used by physicists, chemists, and engineers may be unfamiliar to humanitarian demining personnel, and this further exacerbates a historical tension between the perhaps overly optimistic technologist and the perhaps equally jaded field operator. The key to changing this dynamic is to focus on evaluation of the performance of individual systems by designing good experiments, understanding the metrics of success, and conducting good tests and analysis. This makes for smart consumers of the technology and sets standards of performance that developers understand.

Any detection system, or sensor, can be evaluated by comparing it to statistically random detection events. Notably, virtually any system can serve as a detector. If hundreds of lawn darts are dispersed on a minefield, some of those lawn darts will happen to land over mines. The more darts thrown, the higher the probability that each mine will be 'detected'. If enough darts are thrown, the entire field will be covered and all mines will be "detected." But this is not a useful sensor since it has performance equal only to the statistically random detection expected under such circumstances. The goal is to have detections only where there are mines and no false alarms. Comparisons of sensor performance that include this important interrelationship are made using a receiver-operating characteristic curve.

The ultimate goal of an ROC curve is to determine the ability of the detector to indicate that a mine is present (the probability of detection, P_d) as it relates to the detector's tendency to indicate that there is a mine where there is none (the probability of false alarm, P_{fa}). An ROC curve provides this data over a range of sensor sensitivity settings, and a sensor that has a high P_d with a low P_{fa} is a good sensor.

This is not as complex as it might seem at first, and although the underlying technical theory is well beyond the scope of this paper, a simple example will allow basic evaluations to be conducted in the field by humanitarian demining personnel in real field settings. Field workers should not be at a loss when confronted with a new demining technology. Rather, insist on a well-designed field test that will result in good ROC curve data, and then it is possible to assess the technology at the system level without understanding all the intricate details of the underlying technology itself.

Let's use an example minefield, as shown in Fig. 1.

This notional minefield will have ten mines in it as shown in Fig. 2.

These mines should be buried in random locations using typical emplacement techniques. Any visual surface indications of a mine emplacement should be hidden from view or masked.

Each of the one hundred one-meter squares in the notional minefield is then further subdivided. For our purposes here, we will divide each one-meter square into squares approximately the area covered by a single sensor measurement (a circle ~20 cm in diameter). We'll call this the "sensor footprint." Every sensor

Figure 1.   Diagram of a 'Notional Minefield'

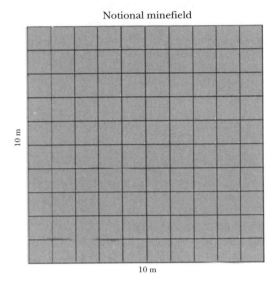

Notional minefield

Field is 10m by 10m, divided into 1m by 1m grid.

Figure 2.   Diagram of a 'Notional Minefield' with Mine Locations Shown

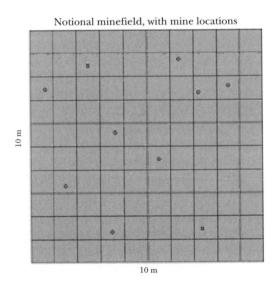

Notional minefield, with mine locations

There are 10 mines buried in the field.

footprint will represent a unique opportunity for a detection or for a false alarm. This is illustrated in Fig. 3.

This means that in the ten-meter by ten-meter minefield, there are one hundred individual one-meter by one-meter grid elements, and each of those grid elements has twenty-five possible unique sensor footprint locations. That makes for a total of 2,500 possible unique readings on the minefield, only ten of which contain mines.

Once an appropriate test area is designed, measurements can begin. Let's assume we are using a metal detector. The sensitivity (or threshold) of the sensor must be kept constant throughout the test, and data should be collected over the entire minefield. Whenever the metal detector indicates that a mine is present, a flag should be placed on the ground. The flag locations should be recorded. The sensor setting should be adjusted a few more times and the experiment repeated. The result on an individual grid element might be as shown in Fig. 4.

The results from this one grid element can be used to calculate P_d and P_{fa} as shown below in Table 1.

Figure 3.   Diagram of One Grid Element in "Notional Minefield"

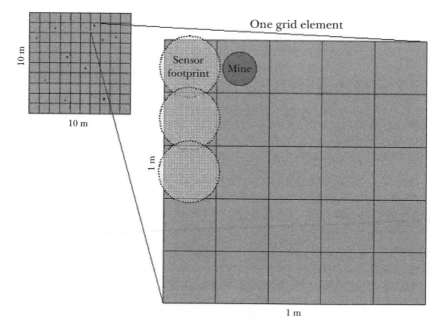

Figure 4. Diagram of Results of Testing on One Grid Element, with Three Different Sensor Sensitivity Settings

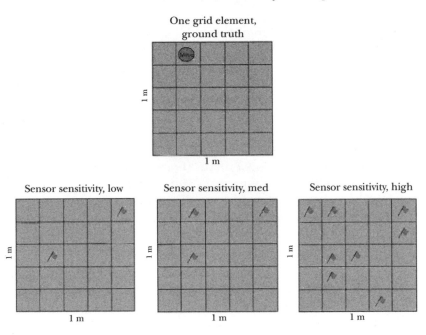

This table of results also illustrates the importance of having a sufficient number of mines in the test to permit statistically relevant results. The results for the entire notional minefield might be as shown in Table 2. (Recall that there are 2,500 unique locations defined by a 20 cm diameter sensor footprint. If 10 of those locations contain actual mines, then the number of false alarm opportunities is 2,500 minus 10 or 2,490.)

These results can be plotted to make an ROC curve for this particular sensor as shown below in Fig. 5.

INTERPRETATION OF THE ROC CURVE

The plot shown in Fig. 5 contains several curves. They are labeled "Random" and "$d = 0.5$" through "$d = 3.0$." These curves are calculated based on the theory governing sensor performance. There are many books describing this theory for the interested reader to investigate further. Suffice it to say that these curves

Table 1.   Results of Sensor Testing on One Grid Element of Notional Minefield

Sensor sensitivity setting	# mines	# mines detected	P_d	# false alarm opportunities	# of false alarms	P_{fa}
Low	1	0	0%	24	1	4%
Med	1	1	100%	24	2	8%
High	1	1	100%	24	7	29%

represent statistically convenient representations of the signals that might result from an actual target, noise, or background clutter. What is important to understand is that when a sensor conforms to this basic theory (usually a good assumption except at the far extremes near $P_d = 0$, and $P_{fa} = 0$ and $P_d = 1.0$ and $P_{fa} = 1.0$), the performance of that sensor over many different sensitivity settings is described by a single curve. In the case of our example, the metal detector would be described by the curve marked "$d = 1.0$" We can then use this curve to determine the performance at any of the various sensitivity settings. Clearly, as the sensitivity is increased, the ability to locate mines increases, but there is a penalty in false alarms.

Note also that ROC curves can be used quite effectively to compare individual sensors. Suppose that we test three different sensors on the same notional minefield. Each sensor experiment is conducted at a single sensitivity level, rather than at three separate sensitivity levels, with the results shown in Table 3.

In the absence of an ROC curve method for comparison, it

Table 2.   Results of Sensor Testing on Entire Notional Minefield

Sensor sensitivity setting	# mines	# mines detected	P_d	# false alarm opportunities	# of false alarms	P_{fa}
Low	10	2	20%	2,490	62	2.5%
Med	10	7	70%	2,490	698	28%
High	10	9	90%	2,490	1,444	58%

Figure 5. Results of Testing in the Form of a Receiver Operating Characteristic (ROC) Curve

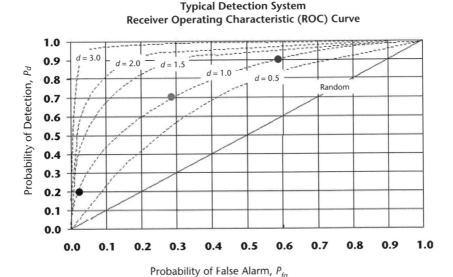

Typical Detection System
Receiver Operating Characteristic (ROC) Curve

Note: This plot is not mathematically precise. It is used for illustration purposes only.

Table 3. Results of Three Different Sensor Tests on Entire Notional Minefield

Sensor	# mines	# mines detected	P_d	# false alarm opportunities	# of false alarms	P_{fa}
Sensor 1	10	4	40%	2,490	448	18%
Sensor 2	10	7	70%	2,490	698	28%
Sensor 3	10	6	60%	2,490	187	7.5%

would be difficult to determine which sensor of the above sensors is the best. It might be tempting to choose sensor 2 with a P_d = 70%, but as shown below in Fig. 6, this would be the wrong choice. Sensor 3 is actually the best sensor of the group.

Indeed, using this analysis, we can compare sensors tested at different levels on equal footing. With an ROC curve analysis, we can estimate the relative performance of each sensor when operated at a sensitivity level that results in location of 80% of the mines (P_d = 80%). Table 4 shows the results of this analysis. Clearly, sensor 3 is the more desirable detector, with almost half the number of false alarms at a P_d equal to 80%.

Figure 6.   Comparisons of Three Different Sensors Operating at Different Points on Their Respective ROC Curves

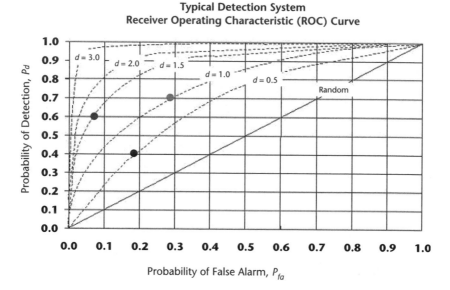

Note: This plot is not mathematically precise. It is used for illustration purposes only.

CONDUCTING REAL TESTS

The above examples are provided only to illustrate the fundamental concepts involved in designing a mine detection test for the field. The actual mines used should be as close as possible to real mines, at least with respect to the attribute that is to be detected. For example, if explosive detection devices are to be tested, the explosive charge should be intact with only the fuse mechanism deactivated or removed. If metal detectors are to be tested, then the mine targets should contain the right amount of metal. Note also that it is critically important not to reveal to test personnel the number of mines buried in the field as even this amount of knowledge can skew test results.

Table 4. Results of Three Different Sensor Tests on Entire Notional Minefield

Sensor	# mines	# mines detected	P_d	# false alarm opportunities	# of false alarms	P_{fa}
Sensor 1	10	8	80%	2,490	1,370	55%
Sensor 2	10	8	80%	2,490	1,058	42.5%
Sensor 3	10	8	80%	2,490	685	27.5%

Further, ten mines is not enough to provide statistically meaningful results; many more mines should be used in an actual test. This has implications for the cost and effort involved in setting up an appropriate test facility. However, doing so is critical to obtaining a realistic evaluation of detection technology and, in the end, saves time, money, and lives.

INFRASTRUCTURE

Even the above tests, while clearly executable by field personnel, would be made easier by the involvement of technical personnel. A fundamental problem in transitioning promising new technologies to the field is that there is currently no infrastructure to facilitate and interpret real technological breakthroughs. As such, the humanitarian demining community is often not able to absorb advances. Little help is currently available to assist in the evaluation of technologies for specific field applications, to bring together the necessary components of a solution, or to assess the real risk to deminers. Indeed, this is in part why historical attempts to use advanced technology in demining operations is peppered with promising starts that falter rapidly. A large infrastructure gap exists that needs to be filled before the technological solutions developed in the laboratory environment can be consistently developed and fielded such that they meet the needs of humanitarian deminers.

BUILDING THE RIGHT INFRASTRUCTURE

The supporting infrastructure should consist of a minimum of four elements:

1. An expert and world renowned "reconnaissance" team
2. A contracted operational capability
3. A technology training and education capability, and
4. A development fund

An expert and world-renowned reconnaissance team consisting of field operation and technology experts is essential for providing independent analyses that match available technologies with specific operational situations. For example, in Cambodia, where the preponderance of the mines are in water-saturated soils, radar techniques would not be suitable. In arid regions, where the soil is highly uniform, radar techniques could provide greater benefit. This reconnaissance team must consist of individuals who do not have a conflict of interest regarding the success or failure of any single technology. In this sense, the team would be able to provide expert advice to indigenous deminers or demining contractors without force-fitting specific technologies. Team members would be able to provide expert assistance in setting up suitable test fields and evaluating test results. In certain situations, the team may decide that the use of an advanced technology is not warranted and in this way ensure "good" technology experiences on the part of the demining community.

A contracted operational element would assemble the best suite of technological solutions for a given demining problem. This provides rapid access to state-of-the-art technologies and keeps the entire entity honest since decisions about the specific solutions then also affect the group making the decisions. This approach would also provide critical technical feedback from the field to technology developers. And perhaps most importantly, it would create a service model that would amortize large capital expense demining equipment over many users.

Another essential element is a training and education site that would allow indigenous capabilities to be developed should an organization or community wish to grow their own.

A modest development fund would allow modifications to be made to existing technologies as needed for individual operations or to initiate the development of missing link technologies. This would ensure that complete solutions are available.

Setting a Goal

The goals of this organization infrastructure would be threefold: (1) to increase the pace of demining worldwide by a factor of ten without sacrificing the safety of local populations or deminers, (2) to serve as the expert system integrator providing complete advanced technology demining solutions to the world, and as such, (3) to serve as a focal point for contributions to demining operations worldwide.

Identifying Supporting Funds

Funding for this organization might be provided from a variety of sources, primarily government, private sector commercial, and philanthropic. The U.S. Department of State has a keen interest in fostering rapid repatriation, stabilization, and economic recovery after regional conflicts. Often the presence of land mines is a limiting factor in these efforts. As such, the State Department has a vested interest in providing the building blocks necessary to ensure that technologies developed for demining find appropriate and rapid application in the field.

Many of the countries where mines limit land use are prime candidates for development of large-scale agriculture. Agricultural product companies cannot currently gain access to these markets because of the problems presented by land mines. In addition, it is well understood that many of the countries with severe and lingering mine problems must rebuild their agricultural capabilities in order to stabilize their economies. An opportunity exists for cooperative work with companies interested in legitimate business growth that also advances the recovery and economic growth of the host country. Such partnerships should be nurtured.

Finally, the donor community has largely turned away from the demining problem in frustration over the painstakingly slow progress that has been made and the complicated network of participants. A nonprofit organization focused on filling the gaps in infrastructure with combined technical and operational expertise would provide the focus necessary for rekindling the philan-thropic participation in global humanitarian demining.

CONCLUSION

The problem of land mines impacts countries, communities, fam-ilies, and lives. Current demining techniques are painstakingly slow and dangerous. The key to improving the pace at which de-mining occurs is to build a strong bridge between humanitarian workers and the technology development community. This bridge has a foundation comprised of two parts: (1) a robust tech-nical and operationally savvy organization that can provide the necessary infrastructure to facilitate the transition and applica-tion of new technologies, and (2) better test design, conduct, and evaluation to allow users of the technology to more realistically assess the performance of any new technology in their field set-ting.

The first will take an investment over two-to-five years to assem-ble the necessary components, but efforts in the second can start immediately.

Such an approach would allow us to begin the process of rid-ding the world of land mines. We would no longer require dec-ades but rather years—not generations of efforts, but the concerted effort of one.

CONCLUSION

Kevin M. Cahill, M.D.

HOSPITALS, once charnel houses where one isolated the incurable and the contagious from contact with healthy society, and which became holding centers to await the inevitability of death, are now centers of research and hope as well as service.

A physician spends his/her day—everyday—surrounded by the suffering of patients. To a degree, depending on the qualities of the varying doctors, the physician actually attempts to experience and share in that suffering so that comprehension of pain becomes an integral part of the healing process. The good physician realizes the privilege of his/her position—to be healthy and fit among the sick and maimed, to possess the powers and medicaments to relieve pain, to offer a cure to the despairing, and offer the glimmer of being able to "go on" to those who have lost all hope.

To fill his/her role successfully the physician must acquire the skill of detachment so that he can direct order out of chaos, arrest and change a fatal course of events. But that detachment need not be cold and selfish; in fact, it must be compassionate and involved so that a proper therapy can be fashioned for the particular needs of the individual patient. The wise physician also knows that his/her every decision can make matters worse. If he/she makes the wrong choice regarding a drug or the timing of surgery and therapy, the patient may die an iatrogenic death. Even while providing correct therapy, doctors can cause extreme pain and suffering—consider the impact of aggressive chemotherapy or radical surgery removing vital organs to preserve the vestiges of life.

Now take the leap—and it is a quantum expansion of pain and suffering—to the situation that faces humanitarian workers following natural disasters and armed conflicts. Enter, in your minds, refugee camps where tens and often hundreds of thousands of the innocent—mostly women, children, and the frail elderly—try to survive, unwanted and confused, hungry and ill, in a devastated land, fearful for their lives and families, vulnerable to the prey of the conqueror or the lawless rabble. Think of the millions of internally displaced persons around the globe, not transient images on a TV screen but real human beings fleeing from oppression and rape and starvation.

This situation is the modern civilized hospital setting turned on its head: the luxury of individual care, time for reflection, and the cultivated niceties of medical service must be sacrificed, at least in the early phases of a complex humanitarian emergency, to the overwhelming need to impose an order and a system that can help save the greatest number of lives.

Here is where technology can complement compassion, can make possible the aspirations of the good physician or skilled humanitarian worker. Without drawing on the technical capabilities—and potential—of the most advanced nations, those caught up in the almost endless maelstrom of humanitarian crises that scar our earth will continue to needlessly suffer and die.

This text offers seeds of hope for a world in desperate need of fresh ideas and solutions. It brings together, within the covers of a book, some of the finest minds in the field of technology and links their efforts, previously largely devoted to critical defense issues, to the problems faced daily by humanitarian workers. Our hope is also that those who read this volume will find their own innovative ways to share in, and help alleviate, the suffering of mankind.

NOTES

INTRODUCTION

1. Carol C. Adelman, "The Privatization of Foreign Aid: Reassessing National Largesse," *Foreign Affairs* 82 (November/December 2003): 9–14.

TECHNOLOGY AND HUMANITARIAN ACTIONS: A HISTORICAL PERSPECTIVE

1. S. R. Roff, *Hotspots: The Legacy of Hiroshima and Nagasaki* (AEC Research Studies, 1995), chapters 1–2.

2. United Nations Economic Committee for Europe, Environmental Performance Review, Ukraine, Nuclear Management, Chapter 4, "Management of Nuclear Safety," 45–65, www.unece.orsl/env/epr/studies/ukraine/welcome.htm, and Keith Baverstock, "Chernobyl and Public Health: The Nuclear Industry Should Fund an International Foundation to Learn from Chernobyl—Editorial," *BMJ*, 316 (March 28, 1998): 952–53.

3. "News in Brief," *The Lancet* (April 13, 2002): 1322.

4. See the *New York Times* articles for 1957: June 10, 29:3, June 24, 25:6, August 16, 29:1, August 25, 62:1, August 28, 25:1, September 1, 28:5, September 1, Iv, 8:4, September 14, 20:1, April 3, 28:4, and other articles.

5. www.who.int/csr/disease/influenza/en/.

6. Quoted in Richard Preston, *The Demon in the Freezer* (New York: Random House, 2002), 57. The account of the eradication of smallpox and the current threat of recombinant smallpox is taken from this acclaimed book.

7. For medical telecommunications and telemedicine, see Victoria Garshnek and Frederick M. Burkle, "Telemedicine Applied to Disaster Medicine and Humanitarian Response: History and Future," in *Proceedings of the 32nd Hawaii International Conference on Systems Sciences, IEEE,* 1999.

8. "Finding Humanity in Humankind," *Canadian Speeches* 16, no. 3 (July-August 2002).

9. W. R. Aykroyd, *The Conquest of Famine* (New York: E. P. Dutton, 1975), 200–203.

10. Jenny Edkins, *Whose Hunger?: Concepts of Famine, Practices of Aid* (Minneapolis: University of Minnesota Press, 2000), 3, 79–80.

11. Robert I. Rotberg and Thomas G. Weiss, *From Massacres to Genocide: The Media, Public Policy, and Humanitarian Crises* (Washington, D.C.: The Brookings Institution, 1996), 18.

12. S. I. Ivashov and V. N. Sablin, "New Technologies in Humanitarian De-mining Operations," *IEEE International Symposium on Technology and Society,* September 6–8, 2000, 101–5.

13. Richard A. Mathew and Ken R. Rutherford, "Banning Landmines in the American Century," *International Journal on World Peace* 16 (1999).

14. Rudi Volti, *Society and Technological Change,* 4th ed. (New York: W. H. Freeman and Co., 2001), 5.

15. Roberta Cohen and Frances M. Deng, *Masses in Flight: The Global Crisis of Internal Displacement* (Washington, D.C.: The Brookings Institution, 1998), 240.

16. David B. Daatrud, Ramina Samii, and Luk N. Van Wassenhove, "UN Joint Logistics Centre: A Coordinated Response to Common Humanitarian Logistics Concerns," *Forced Migration Review* 18 (September 2003): 11–14.

17. Margaret Vikki and Erling Batheim, "Lean Logistics: Delivering Food to Northern Uganda IDPs," *Forced Migration Review* 18 (September 2003): 25–27.

18. Lars Gustavsson, "Humanitarian Logistics: Context and Challenges," *Forced Migration Review* 18 (September 2003): 6–8.

19. H. Wally Lee and Marc Zbinden, "Marrying Logistics and Technology for Effective Relief," *Forced Migration Review* 18 (September 2003): 34–35.

20. "Humanitarian logistics software being designed for Red Cross Red Crescent," press release, May 22, 2002, IFRCRC, and "Humanitarian logistics enters the 21st century," September 4, 2003, www.ifrc.org/docs/news.

Biometrics: Personal ID/Tagging

1. A false negative in personal identification is defined as the percentage of times when the person *is* not authenticated as who she/he claims to be, but is. For a reliable system of identification and authentication, we need to keep the false negatives to a small fraction.

2. A false positive in personal identification is defined as the percentage of times when the person is authenticated as who she/he claims to be but in reality is not. For a robust system of identification and authentication, we need to keep the false positives to a small fraction.

3. A probability of false negative for detecting the desired object is defined as the incidence of missing the detection of a forbidden item or substance when in fact it is present.

4. A probability of false positive for detecting the desired object is defined as the incidence of the detection of a forbidden item or substance when in fact it is not present.

5. A prime example is waiting in long lines at airports.

WIRELESS TELECOMMUNICATIONS

1. See Juan Navas-Sabater, Andrew Dymond, and Niina Juntunen, "Telecommunications and Information Services for the Poor: Toward a Strategy for Universal Access," World Bank Publications, 2002, www .eldis.org/static/DOC10216.htm.

2. In May 1983 the International Telecommunications Union established an Independent Commission for World-Wide Telecommunications Development chaired by Sir Donald Maitland. In January 1985 the commission submitted its report, known as The Missing Link, later to be known as the Maitland Report.

3. See ITU, "World Telecommunication Development Report: Reinventing Telecoms," 2002, www.itu.int/ITU-D/ict/publications/wtdr_02/.

4. Paisley Richardson, "Why the first mile and not the last?" SDdimensions, July 1999.

5. More detail at grouper.ieee.org/groups/802/16/docs/01/80216-01_58r1.pdf.

6. More detail at grouper.ieee.org/groups/802/20/P_Docs/IEEE %20802.20%20PD-04.pdf.

7. Backhaul connections link the outer-regions communications node or network with the primary regional or national network.

8. Information regarding Latvia and Moldova obtained from INET '99 Proceedings, www.isoc.org/isoc/conferences/inet/99/proceedings/4d/4d_2.html.

9. Cylink was acquired by SafeNet in 2003. See www.safenet-inc.com; INET is an Internet networking conference in association with the Internet Society, ISOC (www.isoc.org).

10. Information regarding Nigeria from "Wireless Networking in Africa, Postogna, Fonda, Caesa, Ajayi, and Radicella," *Linux Journal,* no.

56 (December 1998); ICTP training activities in telecommunications, radio communications and computer networking began in 1989 in response to a call for help from developing countries. By 1998, they had held some thirty training sessions attracting 1,372 participants, including 356 from Central and South America, 350 from Europe, 345 from Africa, 321 from Asia, and 31 from North America, Oceania, and international organizations.

11. Information regarding the Solomon Islands obtained at www .peoplefirst.net.sb.

12. See www.schuemperlin.com/ for detail on Wavemail and HF modem and ham.srsab.se/icom_pro/ic78.htm for details on ICOM IC78 HF radios.

13. For example, see Sharp solar power package at sharp-world.com/ corporate/news/030919.html.

14. Information regarding Bhutan from Asia-Pacific Initiatives for the Information Society (AIIS), www.aptsec.org/aiis/AIIS-2/Presentation/ Session%206/6–6/Bhutan%20-%20ict.doc and www.bhutan-notes.com/ clif/; G. O. Ajayi, *Some Aspects of Information Communication Technology Development in Africa*, www.tenet.res.in/commsphere/s8.1.pdf.

15. See www.vocaltec.com/.

16. Mary Greczyn, "VoIp, Wi-Fi Touted for Developing Countries at ITU," *Washington Internet Daily*, October 15, 2003.

17. The amateur radio operators, affectionately called ham operators, were the major exception.

18. Quote from Pat Gelsinger, Intel CTO, "The Wireless Internet Opportunity for Developing Nations," New York, June 27, 2003.

COGNITIVE RADIO FOR HUMANITARIAN OPERATIONS

1. Joseph Mitola III, "Software Radio Architecture: A Mathematical Perspective," *IEEE Journal on Selected Areas in Communications* (New York: IEEE Press, May 1999).

2. Mitola, *Software Radio Architecture* (New York: John Wiley, 2000); W. Tuttlebee, *Software Radio* (London: Wiley Interscience, 2001).

3. Software Defined Radio Forum website, www.sdrforum.org, October 2003.

4. JTRS JPO website, jtrs.army.mil, October 2003.

5. JTRS website, jtrs.army.mil/sections/overview/fset_overview_clusters .html, October 2003.

6. Kendall Grant Clark, "A Web of Rules," O'Reilly XML.com website, www.xml.com, November 3, 2003.

7. IEEE Standard Upper Ontology, www.suo.ieee.org, October 2002.

8. DARPA's XG Program, www.darpa.mil, October 2003.

9. Mitola, "Cognitive Radio for Flexible Mobile Multimedia Communications," *IEEE 1999 Mobile Multimedia Conference (MoMuC)* (New York: IEEE Press, November 1999).

10. Mitola, *Cognitive Radio: Model-based Competence for Software Radios*, (licentiate thesis, Stockholm: KTH, September 1999).

11. Mitola, *Cognitive Radio:An Integrated Agent Architecture for Software Defined Radio* (doctoral dissertation, Stockholm: KTH, June 2000).

12. James Neel et al., "Convergence of Cognitive Radio Networks," WPMC, Blacksburg, Va.: Virginia Polytechnic Institute, 2003.

13. Mitch Kokar, *Ontology-Based Radio*, Boston, Northeastern University, 2003.

14. To address future implementations, one must typically hypothesize designs in a "sequence of implementations" that have not yet been built.

15. Ibrahim Osman, "Scope of International Humanitarian Crises," in *Basics of International Humanitarian Missions*, ed. Kevin Cahill (New York: Fordham University Press, 2003).

16. W. Hayward, "Switching Networks and Traffic Concepts," in *Reference Data For Engineers*, ed. E. Jordan (Indianapolis, Ind.: Howard W. Sams & Co., 1986).

17. Jeffrey Reed, *Software Radio*, Blacksburg, Va.: Virginia Polytechnic Institute, 2001.

18. Jeffrey Steinheider, "Field Trials of an All-Software GSM Basestation," SDR Forum Technology Symposium, Rome, New York, November 2003.

19. FCC Rule and Order, Part 15 Use of Broadcast Television Bands, Washington, DC: Federal Communications Commission, 2003.

20. Although $5.9-2.5 = 3.4$ GHz, only 1.07 GHz of this spectrum is allocated to mobile—hence *sharable*—subbands.

21. J. Mikkonen, *Quality of Service in Radio Access Networks*, Tampere, Finland, Tampere University of Technology, May 1999.

22. Ziemer and Petersen, *Digital Communications and Spread Spectrum Systems*, New York: Macmillan, 1985).

23. T. Finin, "KQML: A Language and Protocol for Knowledge and Information Exchange," International Conference on Building and Sharing of Very Large-Scale Knowledge Bases, Tokyo, December 1993.

24. "UCSD, VA and Cal-(IT)² Wireless Technology To Enhance Mass Casualty Treatment in Disasters," *UCSD Health Sciences News* (San Diego, CA: University of California at San Diego), October 23, 2003.

25. S. Kaufman, *At Home in the Universe*, (New York: John Wiley, 1995).

26. Dorgio and Gambardella, "Ant Colony Systems: Cooperative Learning of the Travelling Salesman," *IEEE Transactions on Evolutionary Computation* (New York: IEEE Press, April 1997).

27. D. Fogel, et al, "Inductive Reasoning and Bounded Rationality Reconsidered," *IEEE Transactions on Evolutionary Computation* (New York: IEEE Press, July 1999).

ENERGY TECHNOLOGIES FOR HUMANITARIAN PURPOSES

1. Vijay Vaoitheeswaran, *Power to the People: How the Coming Energy Revolution will Transform an Industry, Change our Lives, and Maybe Even Save the Planet* (New York: Farrar, Straus and Giroux, 2003).

2. Arif Hepbasli and Zafer Utlu, "Evaluating the Energy Utilization Efficiency of Turkey's Renewable Energy Sources during 2001," *Renewable and Sustainable Energy Reviews* 8 (2004): 237–55.

3. Amulya K. N. Reddy and Jose Goldenberg, "Energy for the Developing World," *Scientific American* (September 1990): 111–18.

4. United Nations 2002 World Summit on Sustainable Development, Johannesburg Plan of Implementation. www.un.org/esa/sustdev/documents/docs.htm; full document at: www.un.org/esa/sustdev/documents/WSSD_POI_PD/English/POIToc.htm.

5. United Nations 2002 World Summit on Sustainable Development, Energy Issues, anes.fiu.edu/Pro/s10Lo.pdf.

6. Joan M. Ogden, "Hydrogen: The Fuel of the Future?" *Physics Today* (April 2002): 69.

7. D. Von Hippel, "Summary Report of the East Asia Energy Futures Project Activities and Accomplishments: 2001 to 2002," Nautilus Institute for Security and Sustainable Development, revised draft, May 14, 2002.

8. "Technological Alternatives to Reduce Acid Gas and Related Emissions from Energy-Sector Activities in Northeast Asia," Nautilus Institute.

9. "Greenhouse Gases from Small Scale Combustion Devices in Developing Countries: Phase II Household Stoves," EPA/600/R-00/052, June 2000.

10. The Smoke Free Turbo Stove by Community Power Corp. is a wood/gas combination stove recently tested at NREL.

11. ORIGO Stove, produced by Dometic AB, Switzerland, is a methanol based household stove.

12. ITDG, "Smoke: The Killer in the Kitchen," www.itdg.org/html/smoke/smoke_report_3.htm.

13. "Refugee Operations and Environmental Management: Selected Lessons Learned," United Nations High Commissioner for Refugees, Engineering and Environmental Services Section, www.unhcr.org/.

14. United Nations High Commissioner for Refugees, Engineering and Environmental Services Division, "Refugee Operations and Environmental Management," www.rmi.org/images/other/Con-UNHCR_SelLesLrn.pdf.

15. solstice.crest.org/discussiongroups/resources/stoves/Miranda/ Ecostove/Ecos tove.html

16. ITDG, "Solar Photovoltaic Energy: Fact Sheet."

17. ITDG technical brief, www.itdg.org/html/technical_enquiries/docs/solar_thermal_energy.pdf.

18. International Directory of Solar Cooking Promoters, solarcooking .org/directory.asp

19. JDA, "Bottles in the Sun," www.sodis.ch/files/SODISReport Uzbekistan_e.pdf.

20. ITDG technical brief, www.itdg.org/html/technical_enquiries/docs/windpumps.pdf.

21. ITDG technical brief, www.itdg.org/html/technical_enquiries/docs/windpumps.pdf.

22. "US-DPRK Village Wind Power Pilot Project," Nautilus Institute, www.nautilus.org/dprkrenew/index.html.

23. U.S. Department of Energy, www.eere.energy.gov/windandhydro/wind_dist_tech.html.

24. U.S. DOE, www.eere.energy.gov/windandhydro.

25. "WINDSTOR" Renewable Energy Generation, Storage and Distribution System, McKenziebay, www.mckenziebay.com/pdf/WindStor PresentationUser-30604–040308.pdf.

26. U.S. DOE Energy Efficiency and Renewable Energy, Wind and Hydropower Technologies Program, www.eere.energy.gov/windandhydro/hydro_advtech.html.

27. ITDG, "Micro-Hydro Power: Technical Brief," www.itdg.org/html/technical_enquiries/docs/micro_hydro_power.pdf.

28. Hydro-eKIDS, Micro hydro power generating equipment, Toshiba, www.atals.com/newtic/data/ekids.pdf.

29. U.S. DOE Energy Efficiency and Renewable Energy, www.eere .energy.gov/windandhydro/hydro_turbine_types.html.

30. U.S. DOE Energy Efficiency and Renewable Energy, www.eere .energy.gov/hydrogenandfuelcells/fuelcells/.

31. U.S. DOE Energy Efficiency and Renewable Energy, www.eere .energy.gov/hydrogenandfuelcells/fuelcells/technical_areas.html.

32. U.S. DOE Energy Efficiency and Renewable Energy, www.eere .energy.gov/hydrogenandfuelcells/fuelcells/stationary_power.html.

33. U.S. DOE Energy Efficiency and Renewable Energy, "The Hydrogen Posture Plan," www.eere.energy.gov/hydrogenandfuelcells/pdfs/hydrogen_posture_plan.pdf.

34. International Energy Agency, "Hydrogen: Today and Tomorrow," www.ieagreen.org.uk/h2rep.htm.

35. Roger H. Charlier, "A Sleeper Awakes: Tidal Current Power," *Renewable and Sustainable Energy Reviews* 7 (2003): 515–29.

36. Practical Ocean Energy Management Systems, Inc., www.poemsinc.org/FAQtidal.html; U.S. DOE Energy Efficiency and Renewable Energy, www.eere.energy.gov/RE/ocean_tidal.html.

Potential Impact of Advanced Vaccine and Water Technology in Humanitarian Operations

1. J. Banchereau, J. Fay, V. Pascual, and A. K. Palucka,. "Dendritic Cells: Controllers of the Immune System and a New Promise for Immunotherapy," in *Novartis Foundation Symposium* 252 (2003): 226; J. A. Berzofsky, J. D. Ahlers, and I. M. Belyakov, "Strategies for Designing and Optimizing New Generation Vaccines, *Nature Reviews Immunology* 1 (2001): 209; O. J. Finn "Cancer Vaccines: Between the Idea and the Reality, *Nature Reviews Immunology* 3 (2003): 630; D. M. Pardoll, "Spinning Molecular Immunology into Successful Immunotherapy, *Nature Reviews Immunology* 2 (2002): 227; H. L. Robinson "New Hope for an AIDS Vaccine," *Nature Reviews Immunology* 2 (2002): 239.

2. H. L. Robinson, "T Cells Versus HIV-1: Fighting Exhaustion as well as Escape," *Nature Reviews Immunology* 4 (2003):12.

3. National Diabetes Statistics, National Diabetes Information Clearinghouse (NDIC), September 25, 2003, diabetes.niddk.nih.gov/dm/pubs/statistics/index.htm#12.

CONTRIBUTORS

Kevin M. Cahill, M.D. is University Professor and Director of The Institute of International Humanitarian Affairs at Fordham University; President and Director of The Center for International Health and Cooperation; Professor and Chairman, Department of Tropical Medicine, Royal College of Surgeons in Ireland; Clinical Professor of Tropical Medicine, New York University; Director, The Tropical Disease Center, Lenox Hill Hospital; and Chief Medical Advisor, Counterterrorism, New York Police Department.

Alan W. Black, Ph.D. is an associate research professor at the Language Technologies Institute at Carnegie Mellon University and Chief Scientist and cofounder of the for-profit spin-off, Cepstral, LLC. He holds a Ph.D. from Edinburgh University and has worked in speech and language processing for some twenty years. His main research topic is speech synthesis, and he is a principal author of the widely used Festival Speech Synthesis System.

Joseph V. Braddock, Ph.D. is Chairman of the U.S. Army Science Board and founder of BDM Federal, Inc., a principal nuclear weapons "failure testing" company in the United States. He is the recipient of the Secretary of Defense Eugene G. Fubini Award, the Defense Nuclear Agency's Exceptional Public Service Award, and Distinguished Service Awards from the Army Science Board and the Association of the U.S. Army.

Joseph Bravman, Ph.D. is a Board Member of the Arthur C. Clarke Foundation, NDIA, the University of Maryland CSHCN, and the Cornell University Engineering School. Bravman was formerly Senior Vice President of Orbital Sciences Corporation, Executive Vice President of Fairchild Space and Defense Corporation, and President of Fairchild's Electronics Systems Division.

Jaime G. Carbonnel, Ph.D. is the Director of the Language Technologies Institute and Alan Newell Professor of Computer Science at Carnegie Mellon University. He holds SB degrees in Math and Physics from MIT and a Ph.D. in Computer Science from Yale University. His research focuses on machine translation, text mining, machine learning, sequence-structure mapping in computational biology, and their various practical applications.

Geoffrey W. Clark, Ph.D. is a professor of humanities in the Imperatore School of Science and Arts at the Stevens Institute of Technology.

Kristina Rodriguez Czuchlewski is a doctoral candidate in the Department of Earth and Environmental Sciences at Columbia University. Her research focuses on the use of satellite and airborne remote sensing technology for natural hazards mapping, response, and recovery.

Regina E. Dugan, Ph.D. is president and CEO, Dugan Ventures, LLC. Prior to founding her own company, Dugan directed a diverse $100 million portfolio of research and development programs at the Defense Advanced Research Projects Agency (DARPA).

Frank L. Fernandez, Ph.D. is a Distinguished Research Professor of Systems Engineering and Technology Management at the Stevens Institute of Technology. Prior to that he was the Director of the Defense Advanced Research Projects Agency in the U.S. Department of Defense.

Ralph B. James, Ph.D. is associate laboratory director for Energy, Environment, and National Security with the U.S. Department of Energy's Brookhaven National Laboratory. He has authored more than three hundred publications, edited eleven books, and holds nine patents. James is the recipient of numerous scientific honors.

Paul J. Kolodzy, Ph.D. is director of the Wireless Network Security Center at the Stevens Institute of Technology. He has also served as Senior Spectrum Policy Advisor at the Federal Communications Commission (FCC) and director of Spectrum Policy Task Force Program Manager at DARPA in the Advanced Technology Office.

Alon Lavie, Ph.D. is an associate research professor in the Language Technologies Institute at Carnegie Mellon University. His main areas of research are machine translation of both text and speech, and spoken language understanding. His current most active research is on the design and development of new approaches to machine translation for languages with limited amounts of data resources.

Arthur Lerner-Lam, Ph.D. is associate director for Seismology, Geology, and Geophysics at the Lamont-Doherty Earth Observatory of Columbia University, and director of the Center for Hazards and Risk Research at the Columbia Earth Institute. His research focuses on the structure and tectonics of plate boundary zones using seismological methods and on issues related to natural hazard risk assessment and management.

Lori Levin, Ph.D. is an associate research professor at the Language Technologies Institute at Carnegie Mellon University. She has a B.A. in linguistics from the University of Pennsylvania and a Ph.D. in linguistics from MIT. Her research focuses on machine translation and other multilingual applications in natural language processing.

Joseph Mitola III, Ph.D. is special assistant, National Security Agency and DARPA Directors' Offices.

C. Kumar N. Patel, Ph.D. is a professor of physics and astronomy as well as a professor of electrical engineering and chemistry, and he was vice chancellor for research at UCLA. He is founder and chairman of the board of Pranalytica, Inc. He received the National Medal of Science from the President of the United States in 1996.

Helen Todosow is a market and information analyst for the Energy, Environment, and National Security Directorate, Brookhaven National Laboratory.

William L. Warren, Ph.D. is managing partner of Sciperio, Inc., and is founder and CEO of VaxDesign Corporation. He is scientific advisor to the National Tissue Emergency Center. He has received numerous international awards for research and development accomplishments, has over 180 technical publications, 7 patents issued, and many patents pending.

Jeffrey G. Weissel is Associate Director for Marine Geology and Geophysics at the Lamont-Doherty Earth Observatory of Columbia University and head of the Remote Sensing and Visualization Laboratory. His research focuses on the use of remote sensing for elucidating earth surface processes and on developing quantitative and predictive methods for characterizing the evolution and erosion of the earth's surface.

Zhen Zhang is a graduate student at the School of Technology Management, the Stevens Institute.

Vignettes

Larry Hollingworth is Humanitarian Programs Director for the Center for International Health and Cooperation. He is also visiting professor in the Institute of International Humanitarian Affairs at Fordham University in New York. He served with the UN High Commission for Refugees in Bosnia and Chechnya, and with the coalition forces and the UN in Iraq. Prior to that he was a British Army officer for thirty years. He is a frequent lecturer on relief and refugee topics in universities and is a commentator on humanitarian issues for the BBC.

Lt. Col. (Ret.) Noel F. Mulliner, OBE, served for thirty-two years as Royal Engineer officer in the British Army. Having spent the last ten years concentrating on humanitarian mine action and the UK contribution to the global effort, he joined the United Nations Mine Action Service (UNMAS), based in New York, in 2000. He is currently the UNMAS technology coordinator and has previously been responsible for coordinating the UN operational support to several mine action programs, including those in Kosovo, Sudan, and Afghanistan.

Nicola "Nicky" Smith is country director, Liberia, for The International Rescue Committee (IRC). She has also worked in complex humanitarian emergency situations in other countries.

THE CENTER FOR INTERNATIONAL HEALTH AND COOPERATION AND THE INSTITUTE FOR INTERNATIONAL HUMANITARIAN AFFAIRS

THE CENTER FOR INTERNATIONAL HEALTH and Cooperation (CIHC) is a public charity founded by a small group of international diplomats and physicians who believed that health and other humanitarian endeavors sometimes provide the only common ground for initiating dialogue, understanding, and cooperation among people and nations shattered by war, civil conflicts, and ethnic violence. The Center has sponsored symposia and published books, including *Silent Witnesses*; *A Framework for Survival: Health, Human Rights, and Humanitarian Assistance in Conflicts and Disasters*; *A Directory of Somali Professionals*; *Clearing the Fields: Solutions to the Land Mine Crisis*; *Preventive Diplomacy*; the new International Humanitarian Books Series of Fordham University Press—*Basics of Humanitarian Missions*; *Emergency Relief Operations*; *Traditions, Values, and Humanitarian Assistance*; *Human Security for All: A Tribute to Sergio Vieira de Mello*, and a standard textbook, *Tropical Medicine: A Clinical Text*, that reflect this philosophy.

The Center and its directors have been deeply involved in trying to alleviate the wounds of war in Somalia and the former Yugoslavia. A CIHC amputee center in northern Somalia was developed as a model for a simple, rapid, inexpensive program that could be replicated in other war zones. In the former Yugoslavia the CIHC was active in prisoner and hostage release and in legal assistance for human and political rights violations, and facilitated discussions between combatants.

The Center directs the International Diploma in Humanitarian Assistance (IDHA) in partnership with Fordham University in New York, the University of Geneva in Switzerland, and the Royal

College of Surgeons in Ireland. It has graduated more than six hundred leaders in the humanitarian world from fifty-three nations, representing all agencies of the United Nations and most non-governmental organizations (NGOs) around the world. The CIHC also cooperates with other centers in offering specialized training courses for humanitarian negotiators and international human rights lawyers. The Center has offered staff support in recent years in crisis management in Iraq, East Timor, Aceh, Kosovo, Palestine, Albania, and other trouble spots.

The Center has been afforded full consultative status at the United Nations. In the United States, it is a fully approved public charity.

The CIHC is closely linked with Fordham University's Institute of International Humanitarian Affairs (IIHA). The Directors of the CIHC serve as the Advisory Board of the Institute. The President of the CIHC is the University Professor and Director of the Institute, and CIHC officer Larry Hollingworth is Humanitarian Programs Director for the Institute.

DIRECTORS

Kevin M. Cahill, M.D. (President)
David Owen
Boutros Boutros-Ghali
Helen Hamlyn
Peter Tarnoff
Jan Eliasson

Peter Hansen
Francis Deng
Joseph A. O'Hare, S.J.
Abdulrahim Abby Farah
Eoin O'Brien, M.D.
Maj. Gen. Tim Cross

INDEX